T0135678

ANTHROPOMETRIC INDIVIDUALIZATION OF HEAD-RELATED TRANSFER FUNCTIONS

ANALYSIS AND MODELING

Von der Fakultät für Elektrotechnik und Informationstechnik der
Rheinisch-Westfälischen Technischen Hochschule Aachen
zur Erlangung des akademischen Grades einer

DOKTORIN DER INGENIEURWISSENSCHAFTEN

genehmigte Dissertation

vorgelegt von

Dipl.-Ing.

Ramona Bomhardt

aus Hess. Lichtenau, Deutschland

Berichter:

Univ.-Prof. Dr.-Ing. Janina Fels
Univ.-Prof. Dr.-Ing. Peter Jax

Tag der mündlichen Prüfung: 10. Juli 2017

Diese Dissertation ist auf den Internetseiten der Hochschulbibliothek online verfügbar.

Ramona Bomhardt

Anthropometric Individualization of Head-Related Transfer Functions Analysis and Modeling

Logos Verlag Berlin GmbH

λογος

Aachener Beiträge zur Akustik

Editors:
Prof. Dr. rer. nat. Michael Vorländer
Prof. Dr.-Ing. Janina Fels
Institute of Technical Acoustics
RWTH Aachen University
52056 Aachen
www.akustik.rwth-aachen.de

Bibliographic information published by the Deutsche Nationalbibliothek

The Deutsche Nationalbibliothek lists this publication in the Deutsche Nationalbibliografie; detailed bibliographic data are available in the Internet at http://dnb.d-nb.de .

D 82 (Diss. RWTH Aachen University, 2017)

ISBN 978-3-8325-4543-7
ISSN 2512-6008
Vol. 28

Logos Verlag Berlin GmbH
Comeniushof, Gubener Str. 47,
D-10243 Berlin
Tel.:   +49 (0)30 / 42 85 10 90
Fax: +49 (0)30 / 42 85 10 92
http://www.logos-verlag.de

Abstract

Human sound localization helps to pay attention to spatially separated speakers using interaural level and time differences as well as angle-dependent monaural spectral cues. In a monophonic teleconference, for instance, it is much more difficult to distinguish between different speakers due to missing binaural cues. Spatial positioning of the speakers by means of binaural reproduction methods using head-related transfer functions (HRTFs) enhances speech comprehension. These HRTFs are influenced by the torso, head and ear geometry as they describe the propagation path of the sound from a source to the ear canal entrance. Through this geometry-dependency, the HRTF is directional and subject-dependent. To enable a sufficient reproduction, individual HRTFs should be used. However, it is tremendously difficult to measure these HRTFs. For this reason this thesis proposes approaches to adapt the HRTFs applying individual anthropometric dimensions of a user. Since localization at low frequencies is mainly influenced by the interaural time difference, two models to adapt this difference are developed and compared with existing models. Furthermore, two approaches to adapt the spectral cues at higher frequencies are studied, improved and compared. Although the localization performance with individualized HRTFs is slightly worse than with individual HRTFs, it is nevertheless still better than with non-individual HRTFs, taking into account the measurement effort.

Zusammenfassung

In einer monophonen Telekonferenz ist es meist schwierig zwischen verschiedenen Sprechern zu unterscheiden, da die interaurale Differenzen und der spektrale Einfluss des Ohres, welche dem Menschen die Lokalisation von räumlich getrennten Schallquellen ermöglichen, fehlen. Somit kann die Sprachverständlichkeit durch die Verwendung von Außenohrübertragungsfunktionen (HRTFs: head-related transfer functions) erhöht werden. Die HRTF beschreibt den Ausbreitungsweg des Schalls von einer Quelle zum Eingang des Gehörs, welcher durch den Torso-, den Kopf- und die Ohrgeometrie beeinflusst wird. Für eine gute binaurale Reproduktion ist es daher erstrebenswert individuelle räumlich hochaufgelöste HRTFs zu verwenden. Dies erfordert allerdings einen erheblichen Messaufwand. Um diesen zu reduzieren, stellt die vorliegende Arbeit Ansätze zur Anpassung von HRTFs auf der Basis von anthropometrische Abmessungen vor. Da die Lokalisierung bei niedrigen Frequenzen vor allem durch die interaurale Zeitdifferenz beeinflusst wird, werden zwei Modelle zur Anpassung dieser Differenz eingeführt und mit bestehenden Modellen verglichen. Darüber hinaus werden zwei Ansätze zur Anpassung des spektralen Einflusses des Ohres bei höheren Frequenzen untersucht, verbessert und verglichen. Die Lokalisationsgenauigkeit mit individualisierten nimmt im Vergleich zu individuellen HRTFs zwar ab, jedoch ist die Lokalisation häufig besser als mit nicht individuellen HRTFs. Daher stellen die vorgestellten Ansätze zur Individualisierung von HRTFs ein Kompromiss zwischen Messaufwand und Lokalisationsgenauigkeit für eine gute binaurale Wiedergabe dar.

Contents

Glossary

Acronyms

CAPZ	Common pole and zero modeling
DR	Dynamic range
DTF	Directional transfer function
FFT	Fast Fourier transform
HpTF	Headphone transfer function
HRTF	Head-related transfer function
ILD	Interaural level difference
ITD	Interaural time difference
JND	Just noticeable difference
LTI	Linear time invariant system
MAA	Minimum audible angle
PC	Principal component
PCA	Principal component analysis
PRTF	Pinna-related transfer function
SH	Spherical harmonics
SNR	Signal-to-noise ratio
TOA	Time of arrival

List of Symbols

f	frequency
H	transfer function in frequency domain
$h(t)$	transfer function in time domain
λ	eigenvalue or frequency
ω	angular velocity with $\omega = 2\pi f$
p	sound pressure
r	radius
ϕ	polar angle

φ	azimuth angle
π	ratio of a circle's circumference to its diameter
θ	polar angle
ϑ	elevation angle
t	time
τ	time delay
$\mathbf{v}$	eigenvector
$\mathbf{V}$	eigenvector matrix
$Y(f)$	transfer function in frequency domain
$X(f)$	excitation function in frequency domain
$y(t)$	response signal of system in time domain
$x(t)$	excitation signal in time domain

Mathematical Operators

arg	argument for complex numbers
arg max	arguments of the maxima
arg min	arguments of the minima
$\mathcal{H}$	Hilbert transform
ln	natural logarithm
log	logarithm
max	maximum of a vector
min	minimum of a vector
∂	partial derivative
σ	standard deviation
$'$	conjugate transpose for matrices with complex entries and the transpose for real entries
Var	variance

1

Introduction

The individualization of the head-related transfer functions (HRTFs) started with the request to improve localization performance in virtual environments without troublesome and time-consuming acoustic measurements. The transfer function characterizes the sound pressure of an incident wave from a source to the ear and can be measured by a loudspeaker and a microphone inside the ear canal (Møller et al., 1995b; Blauert, 1997, pp. 372-373). According to the shape of torso, head and ear, these transfer functions differ for each individual. In general, HRTFs should be measured in an anechoic chamber to reduce the influence of the room which binds the measurement to a specific location. To enable a dynamic virtual environment with head movements, the measurement of an HRTF data set with a high spatial resolution is required.

Individual HRTF data sets provide a better localization performance and lower front-back confusions than non-individual ones (Wenzel et al., 1993). However, the technical effort to measure these HRTF data sets is tremendously difficult (Richter et al., 2016).

Since the HRTF is mainly influenced by the torso, head and ear geometry, the individual HRTFs can be estimated by these anthropometric shapes to reduce the measurement effort and enhance the localization performance compared to non-individual HRTFs. The individual anthropometric dimensions can be measured without special rooms and acoustic equipment.

In this thesis the link between the anthropometric dimensions (Shaw and Teranishi, 1968; Butler and Belendiuk, 1977; Bloom, 1977; Fels et al., 2004; Fels and Vorländer, 2009) and the HRTF data sets to be individualized is the subject of investigation. For this purpose, an anthropometric database with HRTF data sets and the corresponding head geometry of 48 subjects was created (Bomhardt et al., 2016a). These data sets with their associated three-dimensional models of the head and ear provide the basis for developing individualization methods for HRTF data sets (see Chapter 3 and 4).

In particular, localization cues can be categorized as interaural time difference (ITD) at lower frequencies and spectral cues at higher frequencies (cf. Chapter 2, Rayleigh (1907) and Kulkarni et al. (1999)).

The time difference between both ears of an incident wave provides horizontal directional information and is used for localization at lower frequencies. The corresponding wavelengths are longer than the head size and that is why primarily the head width, depth and height influences the individual ITD (Kuhn, 1977). For the adaption of the ITD, two different anthropometric estimation methods of the ITD are proposed and compared in Chapter 5.

In contrast to the ITD, spectral cues provide both azimuth and elevation information (Blauert, 1997, pp. 93-176). These cues are used at higher frequencies due to larger interaural level differences (ILD) and the frequency-dependent directional characteristics of the pinna. Meanwhile the ILD is used for horizontal localization (see Chapter 6), the directional characteristics of the pinna are more important for elevation localization and front-back discrimination. Thus, smaller geometrical shapes of the ear influence localization at higher frequencies. Before going into in detail of the anthropometric individualization, tools to identify characteristic spectral cues for the human sound localization are developed in Chapter 7. Subsequently, two approaches to individualize the HRTF are introduced, enhanced and compared. The first one adapts an existing HRTF data set based on head and ear dimensions (Middlebrooks, 1999a,b), while the second estimates the HRTF data set statistically by ear- and head-dependence (Nishino et al., 2007). For the comparison of both spectral individualization approaches, the previously introduced objective and subjective measures are used.

2

Fundamentals of Human Sound Localization and Binaural Technology

The basic concepts of human sound localization as well as binaural reproduction techniques with head-related and headphone transfer functions are introduced in general in this chapter. Based on the perception thresholds of human sound localization and the binaural reproduction techniques, a summary of existing HRTF individualization methods is given in Chapter 3, and the individualization approaches are developed to further improve the anthropometrically estimated HRTF data sets in Chapters 5, 6 and 7.

This chapter summarizes the perception thresholds of human sound localization and describes HRTFs in detail. Subsequently, reconstruction techniques for HRTFs using orthonormal basis functions such as pole-zeros, spherical harmonics or principle components are explained. These techniques allow a compression of the HRTF data sets as well as the ability to interpolate or individualize HRTF data sets. Finally, binaural reproduction using headphones is discussed.

2.1. Human Sound Localization

One ear enables us to listen to sound sources and specify a rough location of the same. However, two ears enable us to specify the position of the source within a three-dimensional environment more precisely and thereby supplement visual cues. In everyday life, this assists to locate and separate spatial sources which can also be positioned out of the field of vision. This ability will be briefly discussed in the following, starting with the physical propagation of a wave from a sound source.

Assuming that an omnidirectional point source emits a sound wave in the far field, this wave will travel towards the head and its ears. Additionally to the direct sound path, the wave is also de- and refracted by the human body before the wave reaches the ear drums. These physical effects cause delays and attenuation of the arriving signals which are used by the inner auditory system to determine the sound source location.

The following two sections deal with these physical effects which can be split into interaural differences and monaural cues. Additionally, in the third section the role of head movements with regard to localization is explained.

2.1.1. Interaural Time and Level Differences

For almost all positions of sound sources in space, the propagating waves of these sources arrive later at the one ear than at the other. This physical phenomenon results in an interaural time difference (ITD). This difference increases systematically for lateral directions, reaches its maximum and decreases again at the back of the head. The location of the lateral maximum depends on the ear position, and the maximum delay between the ears is approximately $700\,\mu s$ for an adult's head.

The ITD is caused not only by the distance between both ears but also by the de- and refraction of the head. These effects are frequency-dependent whereas Kuhn (1977) explains and approximates the ITD in three different frequency ranges between the lowest and the highest audible frequency: At frequencies below 2 kHz, the head is the major cause of shadowing effects and the reason for the delay of the arriving wave at the averted ear. With increasing frequency the head's diffraction increases due to fact that the wavelength is small compared to the dimension of the head. Consequently, the waves start to creep around the head. The influence of these creeping waves on the ITD, which is especially used for sound localization in the frequency range below 2.5 kHz (Wightman and Kistler, 1992), is therefore limited. Above this frequency, the human auditory system processes the differences between the ongoing fluctuating envelopes in time-domain (McFadden and Pasanen, 1976) which play a minor important role for localization (Macpherson and Middlebrooks, 2002). Furthermore, at higher frequencies, the interaural level difference (ILD) becomes more important and enables an improved sound localization (Rayleigh, 1907; Kulkarni et al., 1999). Besides the ITD, the ILD is frequency-dependent too: It is very small at frequencies below 2 kHz which is due to the large wave lengths in comparison to the head. For shorter wave lengths the attenuation at the averted ear is larger and influenced by the aforementioned creeping waves. The ILD is strongly direction- and frequency-dependent which is why it is studied in Chapter 6 in detail.

2.1.2. Monaural Spectral Cues

Since the ITD and ILD are almost symmetrical to its maximum, there are always two directions featuring the same time difference in the horizontal plane. Therefore, it is almost impossible for humans to distinguish between frontal and rear sources using the ITD and ILD. This phenomenon can also be found for elevated sources on cones around the interaural axis and it is therefore called *Cone of Confusion* (Blauert, 1997, pp. 179-180).

Spectral monaural cues caused by interferences helps the auditory system to localize sound sources on this *Cone of Confusion*. These cues are produced by the fine structure of the pinna at higher frequencies (Shaw and Teranishi, 1968). The first resonance of the outer ear, which is produced by constructive interference, can be observed as a wide direction-independent sound pressure level maximum around 5 kHz in the complete cavum concha. The second mode of the cavum concha has a sound pressure minimum at the crus helix which lies between the antitragus and the lobe (Takemoto et al., 2012). Higher order modes are also influenced by the fossa (cf. Chapter 7). Besides the modes inside the cavum concha, which are almost independent of the direction of the incident wave, destructive interferences, which are caused by the helix and anti-helix, can be observed (Satarzadeh et al., 2007). Dependent on the direction of the incident wave, the distance from these rims to the ear canal differs. Compared to the direct sound the reflected waves from the rims are delayed angle-dependent. Consequently, they produce angle-dependent sound pressure minima above 5 kHz at the ear canal entrance (Lopez-Poveda and Meddis, 1996). Such minima enable the localization on the *Cones of Confusion* (Bloom, 1977).

However, the auditory system is only capable of localizing sources on the cones if the original signal is known (Carlile and Pralong, 1994). Blauert (1969) showed that the perceived sound direction on the cones can be manipulated by the spectrum of the original stimulus: An amplification of a signal in the range below 0.6 kHz or from 3 kHz to 6 kHz provide a frontal auditory event, whereas an amplification from 8 kHz to 10 kHz will indicate an auditory event above the head. If the signal is amplified in the range of 0.6 kHz and 2 kHz or above 10 kHz, the auditory event is perceived in the rear.

2.1.3. Influence of Head Movements on the Localization

Another opportunity to solve the *Cone of Confusion* is to move the head. Due to these head movements, the interaural difference changes and a determination of the direction of the incident wave becomes possible. The widespread *Snapshot Theory* assumes that humans make use of two acoustic images during the movement (Middlebrooks and Green, 1991). A supplement to this theory is a conclusion by Blauert (1997) that information which is obtained by head movements overrides monaural signal characteristics. For lateral sources emitting a stimulus longer than 0.2 ms, the human localization accuracy improves 10 to 15% (Pollack and Rose, 1967). Nevertheless, head movements do not always improve the localization accuracy. Whether they improve or impede the localization depends on the direction of the incident wave, the head movement itself and the stimulus (Middlebrooks and Green, 1991). This quantity of influencing factors makes studies about localization and head movements very challenging.

2.1.4. Perception Thresholds of the Auditory System

While in the previous sections the physical causes and effects of interaural differences and monaural cues are examined, in the current section the focus is on their just noticeable perception thresholds (JNDs). These JNDs help to develop and evaluate the individualization algorithms in Chapters 5, 6 and 7 subjectively. The following insights into different localization methods, gender differences or the influence of the experiment environment, help to design and interpret the results of the listening experiments in Chapters 5 and 7.

The perceivable time and level differences are regarded first. Afterwards, the localization performance for different methods is discussed as well as the influencing factors gender or experiment environment.

Time Differences First of all, the just noticeable ITD change is regarded. Klumpp and Eady (1956) investigated discriminating thresholds of the ITD for frequency-dependent stimuli. The JND for noise was around $9\,\mu s$ between $0.2\,kHz$ to $1.7\,kHz$ and rises for higher frequency ranges. A further study from Zwislocki and Feldman (1956) showed that this JND is also dependent on the sound pressure level. Beside these dependencies, the JND decreases additionally with an increasing duration of the signal and converges for a length of $700\,\mu s$ (Tobias and Schubert, 1959).

Based on the minimum spatial resolution of $2°$ of the human auditory system and an assumed maximum ITD of $790\,\mu s$, Aussal et al. (2012) calculated the JND for a mismatched ITD to $16\,\mu s$. Another study (Simon et al., 2016) used a two-alternative-choice test to determine the JND between two different ITDs. In this case, the subjects had to judge whether the source was located towards the left or right of a presented reference source (Mills, 1958). The resulting JND for oblique frontal directions is approximately $33\,\mu s$ and for lateral directions at approximately $68\,\mu s$. If an ITD is larger than the maximum individual ITD, this leads to a diffuse source which is difficult to localize (Shinn-Cunningham et al., 1998). Such an ITD is called a supernormal cue.

Level Differences Level differences were studied by Mills (1960) in an experiment on dichotic differences that showed a median threshold level around $1\,dB$ at $1\,kHz$. For lower frequencies, the JND is slightly smaller. But above $1\,kHz$, it will drop again to 0.5 dB.

Localization Performance   There are different studies on the localization performance as summarized for example by Blauert (1997, p. 36), Makous and Middlebrooks (1990) or Bahu et al. (2016). As summarized for the perceivable time and level difference, here it also applies that the localization performance depends on the type (impulses, sinusoids, noise or speech) and the sound pressure level of the used stimulus (Blauert, 1997, pp. 39-50).

Early experiments with white noise pulses report smallest audible changes of the incident sounds of $\pm 4°$ for frontal directions, up to $\pm 10°$ at the sides and $\pm 6°$ at the rear (Blauert, 1997, p. 41). The vertical error for speech accounts for $\pm 9°$ in the front, rises to $\pm 22°$ overhead and decreases again to $\pm 15°$ for rear positions (Damaske and Wagener, 1969). Besides the standard deviation, deviations of the perceived from the expected incident direction are observed (Blauert, 1997, pp. 42-50).

Experiments which investigate the human sound localization are often influenced by the applied pointing method due to a non-feasibility of the direct determination of the perceived sound direction (Seeber and Fastl, 2003). In the previously mentioned experiments, identification tasks are used where the subject has to identify the sound by a discrete position. Other possibilities are the verbal report of the position, exocentric or egocentric methods. All these methods have advantages or disadvantages dependent on the tested directions, the pointing accuracy and familiarity of the subject with the method. The verbal report method, where subjects have to indicate the direction verbally to the supervisor, requires trained subjects to obtain accurate results. The same applies to most of the exocentric methods because the subjects have to project the perceived direction of the sound into another coordinate system, for example a sphere where the point of the incident direction has to be marked. In contrast to the egocentric methods, where the subjects have to point in the direction of the sound with respect to their head or body, exocentric methods have greater pointing accuracy in the rear. Although, Mason et al. (2001) report the superiority of the egocentric method in pointing accuracy. Different egocentric methods were used in the past where manual, head, eye, laser and proximal pointing were favored. Especially, the head and eye methods are limited by the locomotor system when body movements are prohibited. The same applies to manual pointing which is very challenging when it comes to positions at the back of the head. Similar to manual pointing is the proximal pointing method. The subjects point relative to the center of their heads with a marker in the perceived sound direction, whereas using this method to point to areas at the back of the head is very difficult due to the limited freedom of movement. Additionally, all egocentric pointing methods

7

show a parallax error.

Not only the identification task takes visual cues into account but also laser pointing methods (Seeber and Fastl, 2003): The subjects have to adjust a laser spot in the perceived direction of sound for this method. As for the eye and head pointing methods, pointing with a laser is restricted on visible directions.

An indirect method of investigating localization performance is the minimum audible angle (MAA) which is defined as the JND in azimuth direction. Early experiments by Mills (1958) showed that this angle depends on the direction and the frequency. For lateral angles the MAA increases from approximately $1°$ at the front to $7°$ at the side for frequencies below $1\,kHz$. If a stimulus provides only high frequencies, especially for lateral incident waves, the MAA can increase to more than $40°$. In contrast to that, the MAA for frequencies above $7\,kHz$ is below $4°$ at the front.

In addition to the experiment design, the stimulus employed is also important as previously stated. Especially the length and onsets can influence localization performance (Tobias and Schubert, 1959; Perrott, 1969; Blauert, 1997, p. 39).

Right Ear Advantage Differences in the hearing sensitivity between both ears are well known. The right ear is commonly about 2 to $3\,dB$ more sensitive than the left ear (McFadden, 1998). This dichotic phenomenon is almost independent of whether the subjects are right- or left-handed. The widely stated reason for this is the specialization of the left cerebral hemisphere for speech as well as the superiority of contralateral ear-cortex pathways (Emmerich et al., 1988).

Gender Several gender differences in hearing performance are reported by Mc-Fadden (1998). For example, the hearing sensitivity above $2\,kHz$ of females is less than for males. Another fact, which is especially interesting for localization tasks, is the advantage of a large head and ears. Due to the fact that males often have larger heads than females, they are in many cases more sensitive to differences in both interaural time and level differences (McFadden, 1998). In addition it is found that males have an advantage in right-monaural vertical sound localization which is probably an effect of the ear size, too (Zündorf et al., 2011).

Influence of the Room Besides the already mentioned factors, human localization is also affected by the experiment room. Most experiments take place in anechoic chambers where no wall influences are present. In case of echoic rooms, their volume, the absorption coefficient of the walls, the positions of the listener and the source will also affect the localization performance.

One of the first experiments on the localization performance under different room acoustic conditions was done by Hartmann (1983). In the experiment laboratory it was possible to change the ceiling height and the wall properties. For the study room, configurations were used where the reverberation time rests between 1 s and 6 s (EN ISO 3382-1). The localization within the room was tested by a loudspeaker identification task in a centered position with a sinus tone. Most subjects performed better in the absorbing room with a low reverberation time. The performance was best when the ceiling was lowered from 11 m to 3.65 m. Hartmann derived a formula which describes a relationship between direct and noise signal energy depending on the source directivity, volume of the room, distance between source and listener as well as reverberation time.

Later, Giguère and Abel (1993) carried out further investigations into an absorbent and a reverberant room where the reverberation time rested between 0.15 and 1.00 s. The subjects were seated in the center of the room and had to identify the playing loudspeaker. In their conclusion, they stated that an increasing reverberation time decreases the accuracy of all frequencies especially for lateral sound source positions and that localization is not solely affected by direct to noise signal energy.

Experiments dealing with the listener position in a room with a low ceiling led to a better localization performance for centered positions (Shinn-Cunningham, 2001; Shinn-Cunningham et al., 2005). Based on these results, it is assumed that wall reflections help to localize the sources at centered listener positions. On the other hand, listener positions close to the wall disturb the interaural time and level difference and lower the localization performance.

2.2. Basics of Signal Processing

Before the physical behavior of the sound pressure in the ear canal can be introduced on a signal theory level, basics of signal processing are explained in this section.

Linear Time Invariant System First of all, for most of the transformations used during this thesis, it is assumed that the described system, which consists of the transfer path from the source to the receiver in the ear canal, is a linear and time invariant system (LTI system) (Ohm and Lüke, 2010, pp. 12-13). If a measurement with real loudspeaker, amplifier, digital-to analog as well as analog-to-digital converter and a microphone is performed in a room, nonlinearities such as distortion are expected. However, in most cases, an LTI system can be assumed.

Therefore, it follows that system responses $y(t)$ can be expressed by the convolution of the excitation signal $x(t)$ and the impulse response $h(t)$ of the system. This impulse response is given by a very brief broadband input signal, a so-called Dirac delta function with an infinitely thin and an infinitely high spike at origin. If the response of a system on an arbitrary broadband signal, the impulse response of the system, and the excitation signal are transformed into the frequency-domain, they can be expressed as follows

$$y(t) \quad = \quad h(t) * x(t) \quad \circ\!\!-\!\!\bullet \quad Y(f) = H(f) \cdot X(f). \tag{2.1}$$

Time-Frequency Relation If the impulse response of a system is to be determined, either such a Dirac delta function or a deconvolution can be used (Oppenheim et al., 1999, pp. 60-61). Due to the fact that the Dirac delta function is impossible to reproduce with real equipment, other broadband signals are often used to measure the system impulse response $h(t)$.

To obtain the system impulse response, the deconvolution is used which is based on the Fourier transform. This Fourier transform provides the link between the time- and frequency-domain so that time signals can be transformed into the frequency-domain and vice versa. The transformation from the frequency into the time-domain is also called inverse Fourier transform (Ohm and Lüke, 2010, p. 67). Since the convolution becomes a product in the frequency-domain, the impulse response $h(t)$ of the system can be calculated from its transfer function $H(f)$ which can be expressed depending on the transfer functions of the input $X(f)$ and output $Y(f)$

$$H(f) \quad = \quad \frac{Y(f)}{X(f)}. \tag{2.2}$$

Subsequently, the transfer function $H(f)$ can be transformed back to the time-domain by the inverse Fourier transform

$$H(f) \quad \bullet\!\!-\!\!\circ \quad f(t). \tag{2.3}$$

For digital signal processing, a discrete sampling is applied which is the sampled version of the continuous measured signal of the system $y(t)$. For acoustic signal processing, sampling rates of $f_S = 44.1\,\text{kHz}$ or $48\,\text{kHz}$ are used. These sampling rates are chosen because of the upper absolute threshold of hearing at approximately $20\,\text{kHz}$ and the Nyquist theorem (Ohm and Lüke, 2010, p. 297). The sampled signals can be transformed into the frequency-domain by the discrete Fourier transform (Ohm and Lüke, 2010, pp. 130-134) which transforms a finite number of equidistant time samples with a spacing of $\Delta t = \frac{1}{f_S}$ into a finite number of frequency bins. To speed up the transformation often the Fast Fourier transform (FFT) is used in computer programs.

Signal-to-Noise Ratio Measured signals are often affected by noise that interferes with the signal of the observed systems. Commonly, the noise is generated by the measurement system itself, for example from the amplifier or quantization noise from the analog-to-digital converter. In addition, the measurement environment can be influenced by interfering external sources. Since in most applications the signal of the observed system is in focus, the mentioned influencing sources should be kept as low as possible.

To indicate if a signal is noisy, the signal-to-noise ratio (SNR) is used

$$\text{SNR}_{dB} = 20\log_{10}\left(\frac{\tilde{p}_{signal}}{\tilde{p}_{noise}}\right). \tag{2.4}$$

Instead of using the ratio of signal power to the noise power, the SNR is already defined for measured effective sound pressure $\tilde{p}$ here.

Alternatively, the dynamic range can be used

$$\text{DR}_{dB} = 20\log_{10}\left(\frac{\max p_{signal}}{p_{noise}}\right) \tag{2.5}$$

where $\max p_{signal}$ is the maximum amplitude of direct sound and p_{noise} the noise floor.

2.3. Head-Related Transfer Functions

The described interaural differences as well as con- and destructive interferences in Section 2.1.2, which can be also summarized as monaural cues, can be monitored by directional, distance and frequency-dependent influences on the spectrum of the perceived signal. The transfer path from the source to a point in the ear canal characterizes the so-called head-related transfer functions (HRTFs) with the acoustic influence of the human body (Møller et al., 1995b; Blauert, 1997, pp. 372-373). Blauert (1997, p. 78) defines these free-field HRTFs as follows

$$\text{HRTF}_{free\text{-}field} = \frac{h_{ear}}{h_{ref}} \tag{2.6}$$

where h_{ear} is the measured transfer function from the source to the microphone in the ear canal and h_{ref} is the transfer function at the position in the center of the head with the subject not being present[1].

If the sound source can be considered as a point source, it can be described by *Green's function* which expresses the sound propagation of a point source

---

[1]The monaural and interaural transfer functions, which are also explained in the book by Blauert (1997, p. 78), are not considered in this thesis. If nothing else is specified in the following, the designation HRTF means the free-field HRTF.

dependent on the distance to the source, the frequency, the speed of sound and volume velocity. Furthermore, *Green's function* describes the fact that the same sound pressure can be observed when the source and receiver positions are exchanged (Kuttruff, 2000, pp. 62-63). This relationship is called the reciprocity theorem which can be applied for example for simulations (Fahy, 1995; Katz, 2001) or measurements (Zotkin et al., 2006) of the HRTFs by positioning the source in the ear canal entrance.

2.3.1. Directional Transfer Functions

The HRTF itself consists of a direction-dependent and independent part

$$\text{HRTF}_{free-field} \quad = \quad \text{DTF} \cdot \text{comm.} \tag{2.7}$$

The directional part is called directional transfer function (DTF) (Middlebrooks and Green, 1990). It is mainly influenced by the reflecting interfering waves on the shoulders and pinnae as the delay time of these waves is dependent on the direction of sound incidence. In contrast to this, the common transfer function comm, which describes the frequency-independent part, is mainly affected by the longitudinal resonance of the ear canal. The DTFs or common transfer functions cannot be determined by a direct measurement, they have to be calculated from an HRTF data set. For this purpose, the surface-weighted arithmetic mean has to be calculated from all magnitudes of the HRTF data set

$$|\text{comm}| \quad = \quad \frac{\sum_{i=1}^{n} w_i \cdot |\text{HRTF}_{free-field,i}|}{\sum_{i=1}^{n} w_i}. \tag{2.8}$$

The resulting common transfer function has unfortunately no phase, so that it has to be estimated. Middlebrooks and Green (1990) rely on the fact that they are regarding an LTI system, and therefore the system can be split into an all-pass system and a minimum-phase system. The all-pass system is represented by a pure delay which is mainly caused by the ITD. They further assume that the sum over the direction-dependent ITD is almost zero and therefore only the minimum phase is reconstructed from the spectrum using the Hilbert transform $\mathcal{H}$ (Oppenheim et al., 1999, pp. 788-789)

$$\phi_{\text{comm}} \quad = \quad \mathcal{H}\{-\ln|\text{comm}|\}. \tag{2.9}$$

Consequently, the DTF can be expressed as

$$\text{DTF} = \frac{\text{HRTF}_{free\text{-}field}}{|\text{comm}|} \cdot e^{-j\phi_{\text{comm}}} \tag{2.10}$$

which includes the direction-dependent HRTF. If it is not further specified, j is the imaginary number in the following.

2.3.2. Pinna-Related Transfer Functions

It is very challenging to assign physical effects of the torso, head and pinna to spectral extrema of an HRTF. The previously mentioned DTFs already reduce the physical influencing factors on the measured transfer functions (no ear canal effects). The pinna-related transfer function (PRTF) goes one step further and considers only pinna-related effects. Therefore the pinna is isolated and the PRTF is determined angle-dependent relative to the position of the receiver in the ear canal entrance. Nevertheless, different studies used different acoustic isolation materials: Algazi et al. (2001b) used a plate which produces reflections, whereas Spagnol and Hiipakka (2011) used a wooden board with a round polycarbonate sheet in the center and Takemoto et al. (2012) used a perfectly matched layer (numeric simulations). As reflecting surfaces will particularly influence measurement positions close to the surface, in the following, the PRTF is described in an absorbent environment. Notwithstanding, the PRTF is subsequently considered as a referenced measurement as the HRTF$_{free\text{-}field}$ in (2.6). So, the measured transfer function H_{ear} is divided by H_{ref} at the microphone position in the ear canal entrance.

2.4. Definition of the Coordinate System and Spatial Sampling

In general, an egocentric coordinate system is used to describe the angle of incidence from which an HRTF is measured. In this thesis, the x-axis is defined from the center between the ear canals into the frontal view direction. The y-axis is the axis from the center to the left ear canal. Consequently, the z-axis points upwards orthogonally to both other axes.

The direction of an HRTF is often described in a spherical coordinate system with the distance r, the polar angle ϑ and the azimuth angle φ. The radius r describes the distance of the source to the center of the head and the polar angle ϑ, which is also called zenith angle, is defined from the zenith downwards. The azimuth angle φ is determined on an orthogonal projection in the xy-plane of the vector which is defined by the source and the center between the ear canals. In this plane it is measured between the x-axis and the projected vector anticlockwise.

Sometimes the elevation θ is used instead of the polar angle which is positively defined from the horizontal plane towards the zenith.

Some studies use an interaural-polar coordinate system (Morimoto and Aokata, 1984; Algazi et al., 2001a) which maps front-back error on the *Cones of Confusion* to the elevation angle. Therefore the azimuth angle is defined along the interaural axis, which is in accordance with the x-axis from the right to the left ear canal, from the frontal direction anticlockwise. The elevation is defined on concentric circles around the interaural axis from the front towards the back of the head.

The measurement of an HRTF data set, which covers, for example, a whole sphere is only possible in a spatial sampling of directions. The most commonly used spatial samplings are equiangular, Gaussian and hyperinterpolation spherical grids around the center of the head (Gardner and Martin, 1994). Similar to the time-domain sampling, the spatial sampling underlies a binding sampling limit which avoids spatial aliasing. This sampling limit is grid-dependent (Driscoll and Healy, 1994), for example for the interpolation and reconstruction of HRTFs which is further described in Section 2.5.2.

2.5. Reconstruction Techniques for Head-Related Transfer Functions

Sometimes it is not feasible to measure high-resolution HRTF data sets or else the data set should be stored using only a few coefficients to reduce the required memory. In this case, decomposition of a data set using orthonormal basis functions can be applied.

In the following, three different approaches using orthonormal basis functions are introduced: The pole-zero decomposition, which provides a physical link to resonances. Subsequently, the spherical harmonics (SHs) decomposition, which can also be used for spatial interpolation and offers a distance transformation. Finally, the principle component (PC) decomposition, which is a more statistical approach and can be used for the individualization by anthropometric dimensions of an HRTF data set. Due to the fact that decomposition is often performed on the magnitude spectrum $|H|$, the phase has to be estimated for which the phase retrieval is also described.

2.5.1. Representation as Poles, Zeros and Residua

Every LTI system, whether physic or numeric, can be represented by its eigen-vectors and frequencies. In room acoustics for example, it is very common to calculate the sound pressure of a shoe box with *Green's function* (Kuttruff, 2000,

p. 70). This solution expresses the pressure by the modes $\mathbf{v}$ and the eigenfrequencies λ of the room. In an abstract sense, the eigenvector $\mathbf{v_i}$ defines the residuum at the position of the source r_0 and the receiver r in the room. The corresponding resonance frequency is represented by the pole λ_i in the transfer function

$$h\left(s\right) \quad = \sum_{i=1}^{n} \frac{\mathbf{v_i}(r)\mathbf{v_i}'(r_0)}{s^2 - \lambda_i^2} \qquad (2.11)$$

which is here considered in the Laplace-domain (Tohyama et al., 1994)[2]. Here, $s = \sigma + j\omega$ is a complex number frequency parameter with real numbers σ and the angular frequency $\omega = 2\pi f$. Due to damping effects the eigenvectors and eigenfrequencies are complex-valued and the Hermitian $\mathbf{v}'$ has to be calculated from $\mathbf{v_i}(r_0)$.

Besides the representation of the addition of rational functions in (2.11), the transfer function can be expressed by the multiplication of poles and zeros (Ohm and Lüke, 2010, pp. 40-42) or common pole and zero modeling (CAPZ) likewise (Haneda et al., 1994; Kulkarni and Colburn, 2004). The CAPZ modeling is often used for room transfer functions but also for binaural representation of HRTF data sets (Haneda et al., 1999) which takes the eigenfrequencies and the damping of the system into account. What all these representations have in common is that the poles are direction-independent. This fact enables the reconstruction of a whole HRTF data set with equal poles for all HRTFs (Haneda et al., 1999). Nevertheless, the zeros and residua are direction-dependent, and at maximum two residua per direction have to be stored.

Vector fitting algorithms for the representation of transfer function data sets provide a solution to determine the complex poles and residua. By this fitting, the complex direction-independent poles and direction-dependent residua of a whole HRTF data set can be determined (Gustavsen and Semlyen, 1999; Deschrijver et al., 2007).

2.5.2. Representation as Spherical Harmonics

An HRTF data set can also be regarded as directivity of the outer ear which has direction-dependent notches, main and side lobes. Moreover, the HRTF can be regarded as an acoustic radiation problem with a source of a specific volume velocity at the entrance of the ear canal, thus fulfilling the reciprocity theorem (cf. Section 2.3). In this case, the Sommerfeld radiation condition provides a solution for the wave equation and enables the representation of this frequency-dependent

---

[2]The function $\mathbf{v_i}$ describes the eigenvector of room mode, therefore Kuttruff (2000, p. 70) and Tohyama et al. (1994) have to consider an additional constant. If an HRTF data set is fitted, this residuum $\mathbf{v_i}(r)\mathbf{v_i}'(r_0)$ already contains this constant, because an empirical instead of a physical eigenvector is considered (Deschrijver et al., 2007).

directivity by the superposition of *spherical harmonics* (SHs) (Duraiswami et al., 2004).

The sound pressure or transfer function H of a source inside the ear canal can be expressed in spherical coordinates (c.f. Section 2.4) as

$$H\left(r, \vartheta, \varphi, k\right) = \sum_{q=0}^{\infty} \sum_{p=-q}^{q} a_{qp}\left(r, k\right) \cdot Y_q^p\left(\vartheta, \varphi\right) \tag{2.12}$$

dependent on wavenumber $k = \frac{2\pi f}{c_0}$, angle (ϑ, φ) and distance r. The complex SH function with the associated *Legendre polynomial* P of order q and degree p is defined as

$$Y_q^p\left(\vartheta, \varphi\right) = (-1)^p \sqrt{\frac{2q+1}{4\pi} \frac{(q-p)!}{(q+p)!}} \cdot P_q^p\left(\cos\vartheta\right) \cdot e^{jk\varphi} \tag{2.13}$$

with the spherical expansion coefficients a_{qp} which are based on the spherical *Hankel function* of the first kind.

Due to the analytic definition of the orthonormal SH functions, an HRTF data set can be represented for arbitrary angles (ϑ, φ) after the decomposition into SH functions. Furthermore, the original data can be transformed using the *Hankel function* with regard to distance (Pollow et al., 2012).

The maximum number of SH coefficients n_{max} used is limited by the grid type and the number of measured directions n_{Dir} to avoid aliasing. Then the maximum number is calculated from

$$n_{max} = \left\lfloor \sqrt{\frac{n_{Dir}}{c}} - 1 \right\rfloor \tag{2.14}$$

whereby $c = 4$ is chosen for an equiangular, $c = 2$ for a Gaussian and $c = 1$ for an hyperinterpolation sampling (Rafaely, 2005).

Beside the spatial interpolation, the SHs can also be applied to smooth the spectrum using a lower order $n < n_{max}$ of SHs during the reconstruction. The smaller the maximum number n, the smoother is the HRTF which diminishes the fine structure of the HRTF (Romigh et al., 2015). On the other hand, the consideration of fewer SH functions and coefficients provides a more compact representation for storing.

2.5.3. Representation as Principle Components

The *Principal Component Analysis* (PCA) reduces the complexity of a data set $\mathbf{H}$ while containing the majority of the information by means of a statistical approach

(Jolliffe, 2002, p. 1). The approach is based on a coordinate system transformation with orthonormal basis functions. These orthonormal basis functions, so-called *principle components* (PCs), are determined from the variance of the data, so that the main information of the data lies in the first components.

Accordingly, the centered data $\hat{\mathbf{H}}$ can be reconstructed again by weighting these principal component matrices $\mathbf{V}$

$$\hat{\mathbf{H}} \;=\; \mathbf{W}\,\mathbf{V}' \tag{2.15}$$

by scores in a matrix $\mathbf{W}$ (Abdi and Williams, 2010). Hereby, the PCs are represented in a $(n \times p)$ matrix of n independent observations and p variables. The orthonormal PCs in the matrix $\mathbf{V}$ are calculated by solving the right eigenvalue problem of the covariance matrix $\mathbf{\Phi}$

$$(\mathbf{\Phi} - \mathbf{\Lambda})\,\mathbf{V} \;=\; \mathbf{0} \quad \text{with} \quad \mathbf{\Phi} = \frac{\hat{\mathbf{H}}'\,\hat{\mathbf{H}}}{n-1} \tag{2.16}$$

where $\mathbf{\Lambda} = \mathrm{diag}\,(\boldsymbol{\lambda})$ is the diagonal matrix of the eigenvalues $\boldsymbol{\lambda}$. Because the mean of $\hat{\mathbf{H}}$ is zero, the variance matrix $\mathbf{\Phi}$ can be expressed by matrix multiplication. The PCs are always ordered by their variance so that the reconstruction can be applied considering the PCs with the largest variances.

The weights $\mathbf{W}$ for the reconstruction are calculated from the transfer functions and the PCs

$$\mathbf{W} \;=\; \hat{\mathbf{H}}\,\mathbf{V}. \tag{2.17}$$

To obtain the original transfer functions $\mathbf{H}$, the mean $\boldsymbol{\mu}$ has to be added

$$\mathbf{h_i} \;=\; \hat{\mathbf{h}}_\mathbf{i} + \boldsymbol{\mu} \quad \text{with} \quad \boldsymbol{\mu} = \frac{1}{n} \sum_{i=1}^{n} \mathbf{h_i}. \tag{2.18}$$

With respect to HRTF data sets, the PCA has different fields of application. If the weights and components are determined angle-dependent, the weights can be fitted by two-dimensional splines and used for a spatial HRTF interpolation afterwards (Chen et al., 1995). For a modulation of an HRTF data set with a low number of coefficients, the magnitude of the spectrum or DTF can be decomposed and reconstructed by a lower number $n < \mathrm{rank}\,(\mathbf{H})$ of PCs (Kistler and Wightman, 1992). The missing phase is often reconstructed by the minimum phase and an additional delay (Kulkarni et al., 1999; Ramos and Tommansini, 2014) (see Section 2.5.4). Early investigations already showed the relationship between the PCs and the anthropometric dimensions (Middlebrooks and Green, 1992). Furthermore, later studies derived the anthropometric weightings using

Regression Analysis (see Chapter 3 for the summary of studies, which are using PCA on HRTF data sets, and Appendix B for further details).

2.5.4. Estimation of the Phase

As already introduced for the DTF in Section 2.3.1, there are several applications where the phase of an original or calculated signal has to be retrieved. This is not only the case for the common part of the DTF but also when reconstruction techniques such as SHs or PCA are used with real-valued transfer functions or magnitudes.

Kulkarni et al. (1999) summarized and compared different phase retrieval methods which are applicable for HRTFs: Minimum phase, minimum phase plus delay, linear phase and reversed phase plus delay[3] reconstruction. Before details of these methods are considered, the phase of the HRTF should first be investigated. The fact that the human HRTF has to be modeled by a minimum phase and delay is caused by the delayed arrival of reflections from the torso and pinna which affects especially the ipsilateral ear (Ziegelwanger and Majdak, 2014). For a minimum phase system (Oppenheim et al., 1999, pp. 287-288) it has to be assumed that the system is time-invariant. Subsequently, the phase can be split into the minimum and continuous phase of the all-pass system

$$\arg\left(H\left(f\right)\right) = \arg\left(H_{min}\left(f\right)\right) + \arg\left(H_{ap}\left(f\right)\right). \tag{2.19}$$

In this case, the minimum phase is determined by (2.9) from the spectrum which alone is not sufficient for human sound localization (Kulkarni et al., 1999; Mehrgardt and Mellert, 1977). Additionally, an all-pass H_{ap} is necessary which adds a supplemental phase and has a magnitude spectrum equal to unity for all frequencies f. For HRTFs, this all-pass transfer function is often modeled as a pure delay

$$H_{ap}\left(f\right) = e^{-j2\pi f \tau_{ap}} \tag{2.20}$$

which can be assumed as a frequency-independent interaural time delay (Kulkarni et al., 1999). Nevertheless, the ITD is still slightly frequency-dependent (Kuhn, 1977) but humans are less sensitive to these frequency-dependent phase variations of the ITD (Breebaart and Kohlrausch, 2001) therefore this fact is often neglected. Furthermore, minor differences between the blocked-ear method and the measurement at the eardrum should be borne in mind if this method is used (cf. Nam et al. (2008), Section 2.3 and 2.6).

---

[3]This reconstruction type was used to show whether the phase has any influence apart from the delay. Therefore, it will not be discussed in the following.

Another approach is to model the phase linearly

$$H_{lin}\left(f\right) = |H\left(f\right)| \cdot e^{-j2\pi\frac{f/f_S}{\tau}} \tag{2.21}$$

with a constantly decreasing and direction-dependent delay τ (Kulkarni et al., 1999). For some applications this assumption is sufficient, (Rasumow et al., 2014) but this phase and the resulting ITD deviate objectively from originally measured ITDs especially for low frequencies $f < 1\,\mathrm{kHz}$. Therefore, Rasumow et al. (2014) proposed a cutoff-frequency $f_c \leq 1.5\,\mathrm{kHz}$: Below this frequency the original and above it the linear phase is used (Rasumow et al., 2014). However, this approach is not suitable for a phase retrieval.

2.6. Binaural Reproduction Using Headphones

Since headphones are often used in virtual acoustic reality, the acoustic influence of these headphones has to be considered.

With headphones the sound is directly played back over the external ear into the ear canal to the ear drum (cf. Figure 2.1). Consequently, if headphones are used for the binaural playback with free-field HRTFs, which already provide information about the human body, the influence of the headphones, external ear and ear canal has to be compensated to simulate free-field listening (Wightman and Kistler, 2005).

Although the ear canal is taken into account in both, the free-field and headphone listening condition, the transfer function of the ear canal is position-dependent. Therefore, the measurement position of the microphone inside the ear canal has to be considered (Chan and Geisler, 1990). In particular, waves that enter the ear canal are reflected in the canal so that standing waves occur above $3\,\mathrm{kHz}$. This sound field inside the ear canal can be approximated by means of a cone with an oblique-positioned ear drum (Hudde and Schmidt, 2009). Two measurement positions are very common: At the ear canal entrance (blocked-ear) and in front of the ear drum (open-ear). While the position at the ear drum is more clearly defined than the entrance of the ear canal, it is very sensitive to noise. Therefore, the blocked-ear method with a microphone at the ear canal entrance is preferred more often; however, the entrance of the ear canal cannot be determined precisely but it can be estimated roughly $7\,\mathrm{mm}$ in front of the ear drum for frequencies below $6\,\mathrm{kHz}$. Assuming that the transfer path in Figure 2.1 for free-field and headphone listening conditions is equivalent (Wightman and Kistler, 2005), the headphone transfer function (HpTF)[4] has to be measured at the same position

---

[4]Furthermore, if the spectrum of the microphone transfer function is not sufficiently flat, it has to be corrected in the HpTF as well.

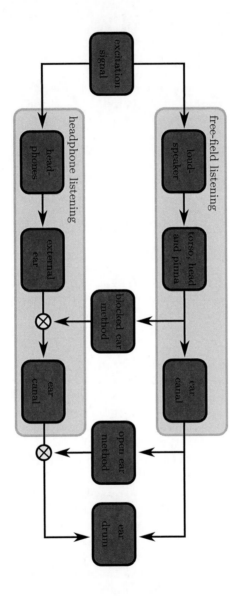

Figure 2.1.: This schematic and approximated representation shows the transfer path from a loudspeaker or headphones to the ear canal (Wightman and Kistler, 2005).

as the HRTF. Consequently, for the free-field reproduction the playback signal has to be multiplied by the HRTF and divided by the inverse of the HpTF in the frequency-domain (Wightman and Kistler, 1989; Møller et al., 1995a). Moreover, for a more realistic scenario the distance has to be considered by using *Green's function*, and a loudspeaker characteristic has to be added.

Anyway, there are different methods to determine the inverse of the HpTF which equalize the headphones and the transfer path of an emitted wave to the ear drum. Møller et al. (1995a) concluded that the average over the sound power considers the peaks better than the level or pressure. Another study proposes to smooth the spectrum of the HpTF by adding twice the standard deviation of eight repeated measurements to their average to provide a smooth and robust equalization with respect to outliers (Masiero and Fels, 2011). Furthermore, a minimum phase for the inverse of the HpTF is supplemented since the auditory system is more sensitive to the spectral cues (Breebaart and Kohlrausch, 2001). In this thesis, the proposed method of Masiero and Fels (2011) is used in the following.

3

Review of Individualization Techniques

If HRTFs are used to create an immersive spatial acoustic virtual reality, the individualization of HRTFs plays an important role. Otherwise, if non-individual HRTFs are used in a virtual scene, the front-back and up-down confusions increase compared to the free-field stimuli, so that the elevation accuracy is often very poor (Wenzel et al., 1993). In other words, the spectral differences, which are mainly used on the *Cones of Confusion* have a significant influence on the localization performance. Additionally, mismatching interaural differences also lead to projection errors in azimuth (Shinn-Cunningham et al., 1998).

Therefore, a general overview of individualization approaches is given in the following in brief paragraphs and will provide a basis for the individualization approaches presented in Chapters 5 and 7. In Chapter 5 approaches to modelling or individualizing the ITD are introduced, discussed and compared with the proposed models of this thesis. The same applies to the individualization approaches of the spectrum in Chapter 7.

Measurement The most accurate approach is the direct measurement of an HRTF data set as described in Chapter 4. Nonetheless, for individual free-field HRTFs special equipment and laboratories are necessary which makes this approach inefficient for commercial use (Møller et al., 1995b; Richter et al., 2016). Different measurement techniques are summarized by Masiero (2012, pp. 21-22) and are also discussed in Section 4.1. To reduce the measurement effort, a small set of measurements can be used to estimate or choose an HRTF data set (Parseihian and Katz, 2012; Xie, 2012; Iida et al., 2014; Maazaoui and Warusfel, 2016).

Averaging Another option is to average several HRTF data sets to obtain a generic HRTF data set. Nevertheless, this approach will smooth the notches of the HRTF which provide important spectral information for localization. Using the HRTF data sets of artificial heads with detailed ear models remains the important spectral fine structure for the localization (Gardner and Martin, 1994; Schmitz and Bietz, 1998) since most of these heads are based on averaged anthropometric dimensions. Due to the fact that the computation time decreases for optimal numeric approaches, the determination of an HRTF data set from an averaged ear geometry is also feasible (Kaneko et al., 2016). Nevertheless, differences between the computed HRTF data set and the individual one remain so that such a data set is not suitable in every case.

Simulation Not only HRTF data sets of an averaged geometry but also of an individual geometry can be calculated numerically. Therefore, either a geometric model, which can be adapted by anthropometric dimensions, or a complete model of the subject's head should be available (Katz, 2001; Fels et al., 2004; Kahana and Nelson, 2007; Dellepiane et al., 2008; Gumerov et al., 2010; Ziegelwanger et al., 2015).

Subjective Selection If an HRTF database is available, HRTF data sets can be chosen subjectively. One strategy is to select a couple of HRTF data sets employing a quick test and to evaluate the remaining data sets in more detail (Seeber and Fastl, 2003; McMullen et al., 2012). Another way is the use of a single-elimination tournament listening experiment design which excludes one HRTF in every single round (Iwaya, 2006; Honda et al., 2007). Simple rating strategies with scales or just with a *good* or *bad* rating can also be used for the selection (Katz and Parseihian, 2012; Parseihian and Katz, 2012). Otherwise criteria such as perceived direction or distance, front-back confusions and externalization can be evaluated to select the most suitable HRTF data set.

Objective Selection Similar to the subjective selection, the best fitting HRTF can also be chosen using anthropometric data (Zotkin et al., 2003; Schönstein and Katz, 2010; Torres-Gallegos et al., 2015). A database, which provides anthropometric data and corresponding HRTF data sets is necessary. Another promising strategy is to measure HRTFs from different directions in rooms and use those to select a data set by minimizing the discrepancy between the measured and the previously recorded (Iida et al., 2014; Maazaoui and Warusfel, 2016).

Tuning For the active sensory tuning, parameters such as poles and zeros of a transfer function are adapted by an expert reflecting the response of the subject (Runkle et al., 2000). After several optimization steps, the HRTF data is fitted. But also parameters such as coloration or spatial resolution can be tuned using equalization, smoothing and phase adaption (Silzle, 2002). Tuning can also be applied on principal components to customize the HRTF (Hwang et al., 2008; Shin and Park, 2008; Fink and Ray, 2015).

Physical Features The relationship between physical dimensions of the human head and HRTFs is well known (Shaw and Teranishi, 1968; Butler and Belendiuk, 1977; Bloom, 1977; Fels and Vorländer, 2009). Either analytic geometry models for a direct calculation or HRTF manipulation on the basis of anthropometric dimensions can be used. The frequency scaling is one very efficient way to individualize HRTFs (Middlebrooks, 1999b). An optimal scaling coefficient can be found by minimizing the inter-subject difference between two HRTF data sets. Applying this optimization to a whole database, the optimal scaling coefficient can be expressed by anthropometric dimensions. Additionally, a rotation shift can be applied to consider the shape and orientation of ear and pinna (Guillon et al., 2008). The principal component analysis in combination with a regression analysis provides another opportunity to calculate individualized HRTFs. In this case, the weights of the principal component analysis are expressed by a linear combination of anthropometric dimensions (Kistler and Wightman, 1992; Jin et al., 2000; Inoue et al., 2005; Nishino et al., 2007; Hugeng and Gunawan, 2010; Ramos and Tommansini, 2014). Both approaches are statistical approaches and need initially a database of HRTF data sets with ear dimensions.

Different analytic and numeric geometry models are known to calculate HRTFs or features of them. Most of them describe the ITD using a spherical or elliptical approach (for example, Woodworth (1940), Kuhn (1977) or Duda and Algazi (1999)). Also models of the torso and head are used to calculate the HRTFs in the frequency-domain (for example, Sottek and Genuit (1999) or Algazi et al. (2001a)). For higher frequencies ear models which describe the geometry of the cavum concha are available (for example, Lopez-Poveda and Meddis (1996) or Spagnol and Geromazzo (2010)).

Characteristic HRTFs As introduced in subjective selection approaches, large databases can be used to group non-individual HRTFs by subjective evaluation (Katz and Parseihian, 2012) or physical features. Moreover, spectral differences of HRTF data sets can be used for clustering (Wightman and Kistler, 1993; Xu et al., 2008). Due to the grouping by characteristic features, a number of data sets can be reduced for the selection process.

4

Individual Head-Related Transfer Functions, Head and Ear Dimensions

For a detailed study of the relationship between the anthropometric dimensions and the corresponding HRTF data sets, a large database which provides both is necessary. In this chapter, existing databases with HRTF data sets and anthropometric dimensions are summarized first. Subsequently, the database, which is used to investigate the individual anthropometric dimensions and HRTF data sets, is introduced.

Summary of Available Databases One of the first freely-available databases was created by the Center for Image Processing and Integrated Computing (CIPIC) at Davis University of California and provides data of 45 subjects (Algazi et al., 2001c). The HRTFs were measured in a simple room at a distance of 1.95 m. The microphones were placed at the entrance of the blocked-ear canal of the seated subject. In total 1250 directions were measured in an interaural-polar coordinate system. The anthropometric dimensions of this database, which consider the body, head and pinna, were defined and taken from two-dimensional images. Detailed information can be found in the Table 4.1. Another database (LISTEN) with 51 subjects was established at the same time (Warusfel, 2002). The HRTFs were measured in a full anechoic chamber and have a spatial resolution of $15° \times 15°$. Head and pinna dimensions were documented according to the specifications of the CIPIC database. A larger HRTF database (RIEC) consisting of 105 subjects with 37 head models can be found at the Advanced Acoustic Information Systems Laboratory, Research Institute of Electrical Communication, Tohoku University (Watanabe et al., 2014). The transfer functions were measured in a full anechoic chamber where the subjects were seated with microphones in their blocked-ears. The transfer functions of the data sets were measured in a resolution of $5° \times 10°$. The largest current database is the ARI database (Majdak et al., 2013) which provides data for over 120 subjects. An HRTF data set of this database has a general resolution of $5° \times 5°$ whereas frontal directions are spatially sampled in $2.5°$ steps. The subjects were seated in a semi anechoic chamber and wore

Name	Subj.	Coordinates	Samples	Room	Dist.	Anthrop. dimensions
CIPIC	45	Interaural-polar coord. (1250 dir.): Az. res. 5° from −45° to 45° (additional sampling points at −80°, −65°, −55°, 55°, 65° and 80°), el. res. 5.625° from −45° to 230.625°	200, 44.1 kHz	Simple room with absorbent material on the walls	1.00 m	Dimensions of the body, head and pinna of 43 subjects
LISTEN	51	Spherical coord. (187 dir.): Az. res. 15°, el. res. 15° from −45° to 90°	512, 44.1 kHz	Anechoic chamber	1.95 m	Dimensions of the body, head and pinna
RIEC	105	Spherical coord. (865 dir.): Az. res. 5°, el. res. 10° from −30° to 90°	512, 48 kHz	Anechoic chamber	1.50 m	Scans of the head and torso of 37 subjects
SADIE	18	Ambisonic 5th order (170 dir.)	256, 48 kHz	Anechoic chamber	1.50 m	No data
ARI	120	Spherical coord. (1550 dir.): Az. res. 2.5° within azimuth range of ±45° and 5° otherwise, el. res. 5° from −30° to 80°	256, 48 kHz	Semi-anechoic chamber	1.20 m	Dimensions of the body, head and pinna of 50 subjects.
SYMARE	10/61	Spherical coord. (393 dir.): Az. res. varies elevation-dependent between 10° and 45°, el. res. 10° from −40° and 90° (additional sampling point at −45°)	256, 48 kHz	Anechoic chamber	1.00 m	Dimensions and models of the body, head and pinna (freely available for 10 subjects)
ITA	48	Spherical coord. (2304 dir.): Az. res. 5°, el. res. 5° from −66° to 90°	256, 44.1 kHz	Semi-anechoic chamber	1.20 m	Dimensions of the head and pinna and ear models

Table 4.1.: Seven freely available HRTF databases are summarized in this table by the number of subjects, the spatial sampling points, the number of samples as well as the sampling frequency, the type of measurement room and the available anthropometric dimensions.

the microphones in their blocked-ears. The anthropometric data are provided according to the CIPIC specifications for 50 of 120 subjects. The SADIE database is focused on the demands of Ambisonics and considers 170 transfer functions on a non-equidistant grid so that it is not appropriate for the studies concerning the relationship between HRTFs and anthropometric dimensions (Kearney, 2015). The HRTF database SYMARE (Jin et al., 2014) is provided by the University of Sydney. The detailed torso and head scans of 10 subjects are freely available. In total, the HRTFs of 61 subjects were measured in an anechoic chamber with an elevation resolution of $10°$ between $-40°$ and $90°$ and a variable azimuth resolution at a 1 m distance. The processed HRTFs consist of 256 samples using a sampling frequency of 48 kHz. Additionally, detailed scans of the torso, head and ear are available.

Different databases, which require a request to the authors, are available (see Table 4.2). For example the AUDIS database which provides the HRTF data sets of 20 subjects with 122 directions. The elevation resolution is $10°$ from $-10°$ to $90°$ and the azimuth resolution is $15°$. The blocked-ear HRTFs were measured in an anechoic chamber at a distance of 2.4 m to the seated subjects. The processed HRTFs have 132 samples and a sampling rate of 44.1 kHz (Blauert et al., 1998). The FIU database provides the HRTF data sets and three-dimensional ear scans of 15 subjects (Gupta et al., 2010). Six elevation angles between $54°$ and $30°$ and 12 azimuth directions were measured in a simple room using the blocked-ear method. The sampling rate used was 96 kHz, and an impulse response consists of 256 samples.

A collection of HRTFs can be found at the Takeda Laboratory at Nagoya University (Nishino et al., 2007). The 111 HRTF data sets are measured in a room with a low reverberation time at a distance of 1.52 m in the horizontal plane ($5°$ resolution). The microphones did not block the ear canal completely. The 512 samples long impulse response have a sampling rate of 48 kHz. In addition, the physical ear dimensions of 80 subjects were measured.

Besides the HRTF databases, some research facilities provide the HRTF data sets of artificial heads such as the KEMAR manikin (Gardner and Martin, 1994; Zhong and Xie, 2013b). These artificial heads often have averaged anthropometric dimensions and have the advantage that HRTF measurements are easier because they do not move during measurement.

Database at Hand   In contrast to the mentioned databases, for the study of HRTF features and anthropometric dimensions a database is required which provides spatial high-resolution HRTF data sets with corresponding detailed ear models. Therefore 48 subjects aged 29 ± 5 years were measured. Most of the subjects (35 of 48) were male and Europeans (43 of 48). Three subjects were

Name	Subj.	Coordinates	Samples	Room	Dist.	Anthrop. dimensions
AUDIS	20	Spherical coord. (122 dir.): Az. res. 15°, el. res. 10° from −10° to 90° (some measurement points were omitted)	132, 44.1 kHz	Anechoic chamber	2.40 m	No data
FIU	15	Spherical coord. (72 dir.): Az. res. 30°, el. angles −36°, −18°, −0°, 18°, 38° and 54°	256, 96 kHz	Simple room	No data	Dimensions of the pinna and ear models
Takeda	111	Spherical coord. (72 dir.): Az. res. 5°, el. angle 90°	512, 48 kHz	Low rev. room	1.52 m	Dimensions of the head and pinna of 80 subjects

Table 4.2.: Three HRTF databases (author request) are summarized in this table according to the number of subjects, the spatial sampling points, the number of samples as well as the sampling frequency, the type of measurement room and the available anthropometric dimensions.

Asians and two were South Americans.

The measurement setup as well as the signal processing of the HRTF data sets are described in the following section and by Bomhardt et al. (2016a). Then, the generation of three-dimensional ear models from the magnetic resonance imaging (MRI) scans as well as the resulting ear dimensions according to the CIPIC specifications (Algazi et al., 2001c) are described.

4.1. Measurement of Head-Related Transfer Functions

The ability of the subject to sit or stand still during the measurement setup limits the resolution of the HRTF data set for a single loudspeaker setup. A higher resolution can be achieved using a multiple loudspeaker setup or a reciprocal measurement setup with microphone arrays, for instance Majdak et al. (2007) or Zotkin et al. (2006). A drawback to using microphone arrays is that the source has to be positioned inside the ear canal. Consequently, to avoid a hearing loss on the part of the subject, the signal level has to be low. A multiple loudspeaker setup, which was used for measured individual HRTF data sets, provided a larger SNR and is therefore used for the database at hand.

Measurement Setup   The measurement setup used was designed by the Institute of Technical Acoustics (ITA) at RWTH Aachen University providing a high-resolution and individual HRTF measurement in a very short time period (Masiero et al., 2011; Richter et al., 2016). To reduce the acoustic influence of the measurement aperture, which is shown in Fig. 4.1, the setup had to be built in a filigree manner. In the vertically aligned continuous arc, which provided the cavity for the loudspeakers, 64 loudspeakers were installed. They were placed in polar direction in a resolution of 2.5° on a semi-circle starting at 1.55° and ending at 160°. The radius of the arc and the distance to the center, where the reference transfer functions H_{ref} were measured and the subject was aligned, was 1.2 m which was a trade-off between measurement SNR and far field conditions of the measurement.

Measurements took place in a semi-anechoic chamber with a stone floor which reflects incident waves. The center of the vertical arc was set at 2 m above the floor where the transfer functions of the subjects H_{ear} were measured.

Prior to the start of the measurement, two Sennheiser KE3 microphones, supported by a dome which blocks the ear (Møller et al., 1995b), were positioned at the beginning of the ear canal. Subsequently, the subjects were aligned using a cross-laser. Additionally, a neck support minimized the head-movements during the measurement. For the spherical sampling, the turntable was moved in discrete steps and performs a full rotation of 360° in 5° steps.

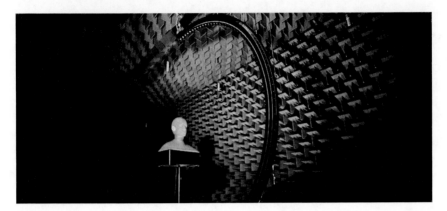

Figure 4.1.: Measurement setup of the arc and an artificial head on a rotating turn table at ITA, RWTH Aachen University.

Measurement Speed-Up This measurement setup with the vertical arc and the rotating subject generated an equiangular grid. For a Gaussian or hyperinterpolation spherical grid either a single movable loudspeaker or fixed loudspeakers have to be used. In this manner, the measurement time or the required equipment will increase for both grids.

In recent years, several methods to further improve measurement speed were proposed using multiple speaker arrays (Majdak et al., 2007). In the present setup the measurement speed improvement was achieved using an optimized multiple-exponential sweep method (Dietrich, 2013). This method uses exponential sweeps with a delay between speakers that is much shorter than the sweep length. The total measurement time for a $5° \times 5°$ measurement data set, where every second loudspeaker of the arc was used, amounted to 6 minutes.

Signal Processing Next, the free-field HRTF data sets were calculated from the measured transfer functions according to (2.6). Undesired reflections on the stone floor in the semi-anechoic chamber had to be removed for free-field compensated HRTFs. In Fig. 4.2 it can be observed that the first reflections arrived 5 ms after the direct sound. To suppress them, a 10th order Hann window was applied which dropped from 5.0 to 5.8 ms. The total length of the signal was cropped to 256 samples which is desirable for real time applications. Senova et al. (2002) as well as Wightman and Kistler (2005) showed that less than 256 samples (sampling rate of $f_S = 44.1$ kHz) gradually decrease the localization performance of virtual images. In practice, transfer functions with less than 100 samples still provide sufficient representation of virtual sound image localization (Xie and

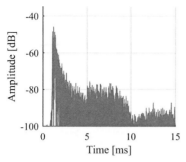

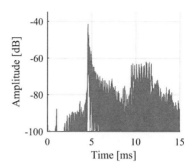

(a) Measured impulse responses h_{ear} at the entrance of the human ear.

(b) Measured reference impulse response h_{ref} at the center of the arc.

Figure 4.2.: The amplitudes of the impulse responses are plotted for every second loudspeaker of the measurement arc for an azimuth angle of $0°$.

Zhang, 2010).

A 1" loudspeaker covered the frequency range between 0.2 and 20 kHz. Nevertheless, the loudspeakers showed a loss of energy at their resonance at 13 kHz which can be well observed in the measured reference transfer function in Fig. 4.3. To suppress angle-dependent notches from the loudspeakers and enhance the SNR of the final HRTFs, the reference transfer functions were regularized (Kirkeby regularization (Kirkeby et al., 1998) with the regularization parameter $\beta = 10^{-10}$ within the frequency range and $\beta = 1$ otherwise). Due to the regularization, the frequency range was limited to 0.2 to 18 kHz.

Despite the careful alignment, it was impossible to position the subject as accurately as an object (Zhong and Xie, 2013b; Andreopoulou et al., 2013, 2015) and the subjects moved their heads during the measurement. Therefore, small angle-independent deviations for the interaural time and level difference can be observed due to an initial misalignment. Analyzing the interaural time difference in the horizontal plane, the angle-independent rotational misalignment rested at $\Delta\varphi_{Off} = -1° \pm 3°$. A detailed description of the correction can be found in Section 5.2 and Bomhardt and Fels (2014). If the sum over the interaural level difference in the horizontal plane is non-equal to zero, this implies either a displacement or an asymmetry of the subject. The level offset $\Delta L = 1 \pm 4\,\mathrm{dB}$ was primarily forced by the displacement and therefore corrected (cf. Section 6.1 for details).

33

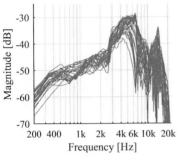
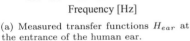
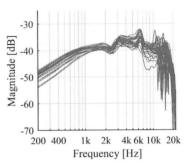

(a) Measured transfer functions H_{ear} at the entrance of the human ear.

(b) Measured reference transfer functions H_{ref} at the center of the arc.

Figure 4.3.: The magnitude of the measured transfer functions are plotted for every second loudspeaker of the measurement arc for an azimuth angle of $0°$.

4.2. Anthropometric Dimensions of the Head and Ear

In contrast to most of the databases in Table 4.1, the head and ear dimensions of the database at hand were obtained from MRI scans and three-dimensional ear models according to the specifications of the CIPIC database (see Fig. 4.4). This has the advantage that hidden surfaces of a three-dimensional model, as for instance at the cymba concha, which is challenging for an optical scanner, are easier to reconstruct.

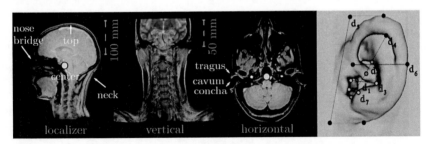

Figure 4.4.: The measurement points for the head width w (tragus & center), depth d_F (nose bridge & center) as well as d_B (neck & center) and height h (top & center) are sketched in the MRIs of subject #17. The ear dimensions according to the CIPIC specification are shown.

4.2.1. Three-Dimensional Ear Models

All of the 48 subjects, who were measured in the HRTF arc, were also measured in an MRI scanner (Siemens MAGNETOM Verio system: Radiologische, Nuklearmedizinische und Strahlentherapeutische Gemeinschaftspraxis in Aachen, Germany). At first, this system scanned the fixed head and generated the so-called localizer scans as in Fig. 4.4.

Figure 4.5.: The vertical MRI scans of subject #17 are stacked up and provide the basis for the three-dimensional ear model.

In a second scan, vertical and horizontal images were taken. From the horizontal scans, which had the highest resolution with a spacing of 1 mm, the ear models were generated. The number of required slices varied dependent on the ear size of the subject.

The scans, as observable in Fig. 4.5, were affected by measurement noise, and the segmentation and geometric reconstruction could not be done immediately from the scans. In a first step, a Gaussian blur filter was applied to reduce the noise. Afterwards, a single layer surface was extracted which still contains noise particles, rough surface areas and gaps. These effects were undesired and forced the following steps: The rough surfaces of the mesh were smoothed using Laplacian smoothing. The remaining non-connected surface parts were removed by filling the holes. Some of the holes could not be filled automatically so that these mesh errors had to be corrected manually. Finally, the number of polygons was decimated which additionally smoothed the mesh[1].

---

[1]This works was done in cooperation with the Chair of Medical Engineering, RWTH Aachen University.

4.2.2. Individual Dimensions of the Head and Ear

The ear models were now usable for simulations but still not sufficient for a geometry analysis or an HRTF estimation. This in particular forced the extraction of the ear dimensions (see Fig. 4.4 with the dimensions according to the CIPIC specifications). In contrast to the extraction of these dimensions from a subject, the three-dimensional ear models feature the advantage of rims and cavities being more easily accessible. Measuring the dimensions from images has the disadvantage that the dimensions vary up to 5 % depending on the perspective (Braren, 2016).

The head dimensions width, depth and height were taken from the MRI scans directly as in Fig. 4.4. The horizontal scan, where the ear canal was visible, was chosen to measure the head width. The head width was taken as half of the smallest width of the head. These measurement points were often located at the tragus of the ear. The axis from the right to the left ear canal was used to determine the height and depth of the head in the localizer scan. From the center of the head between the ears, the depth was measured to the closest point at the front and back of the head. These two points were located at the bridge of the nose and at the neck. The height was determined from this center to the top of the head.

The head and ear dimensions of all 48 subjects are summarized in Table 4.3. The largest dimension was the height of the head followed by the other head dimensions. The ear offset can be determined from the difference between the frontal and rear depth of the head $d_U = d_F - d_B$. Additionally, an averaged depth of the head can be calculated $d_M = \frac{d_F + d_B}{2}$. The largest ear dimensions were the ear height d_5 and width d_6. Both dimensions varied less than the smaller dimensions of the cavum concha and fossa.

	w	d_F	d_B	h	d_1	d_2	d_3	d_4	d_5	d_6	d_7	d_8
Mean μ	71	104	86	133	17	9	18	19	64	36	6	14
Std σ	3	6	7	6	2	2	3	3	5	3	1	2
Min	62	90	70	121	13	5	13	13	53	30	4	11
Max	77	114	107	145	22	11	28	25	74	43	10	19
$100 \cdot \frac{2\sigma}{\mu}$	8	12	16	9	24	44	33	32	16	17	33	29

Table 4.3.: The statistical evaluation of the anthropometric dimensions in millimeters according to the CIPIC specifications and their percentage deviations in %.

In comparison with the CIPIC dimensions, the head width and depth are very similar (deviation is less than 5%), while the height deviates due to its different definition[2]. The cavum concha depth d_8 (37%), cymba concha height d_2 (32%) and fossa height d_4 (26%) from the three-dimensional ear models differ from the CIPIC dimensions too. It is assumed that these dimensions are easier accessible from a three-dimensional ear model as these dimensions are very small in comparison with others. All other dimensions are in comparable ranges with a deviation below 15%.

The correlation between different dimensions is interesting for the anthropometric reconstruction of an HRTF data set in Chapter 7. If two dimensions are highly correlated, one of these two dimensions can be later neglected for the anthropometric estimation of an HRTF data set. Analyzing the correlation coefficient between two parameters, it was found that wider heads often have also a larger depth (correlation coefficient $\rho \approx 0.65$). Most of the ear dimensions (d_1, d_2, d_4, d_6 and d_M) were correlated with the height d_5 of the pinna ($\rho \approx 0.6$), but also a correlation between the fossa height and the pinna width could be observed ($\rho \approx 0.56$). Additionally, the cavum concha height was correlated with the depth and width of the head ($\rho \approx 0.5$).

---

[2]The height is measured from the chin to the top of the head. Halving this dimension provides deviations due to the fact that the dimension from the ear canal to the top is larger than the dimension from the ear canal to the chin.

5

Interaural Time Difference

Apart from the subjective selection of an HRTF data set from a database, the calculation of an individualized HRTF data set by anthropometric dimensions provides a fast and efficient way. Some of the approaches presented in Chapter 3 determine only the real-valued magnitudes and neglect the phase of the complex-valued spectrum of the data sets. In this case, the reconstruction of the phase by a minimum phase and a delay is mandatory (cf. Section 2.5.4). Since the minimum phase is calculated from the magnitude, the delay corresponds to the ITD and can be estimated from anthropometric dimensions.

This chapter introduces two different ITD models. One is based on the analytic solution of the sound pressure on a sphere (Bomhardt and Fels, 2014; Bomhardt et al., 2016b) and the other one is based on the empiric analysis of measured ITDs with the corresponding head geometry. Furthermore, approaches are presented to adapt the ITD of an existing HRTF data set. Both, the models and the adaption approaches, are analytically compared with state of the art models. Additionally, the models presented are subjectively discussed with regard to the just noticeable ITD error.

5.1. Ellipsoid Model

According to the long wave lengths, the head is often approximated as a sphere at frequencies below $1.5\,\mathrm{kHz}$ (Kuhn, 1977). The transfer functions from a source to a receiver on a sphere can be derived analytically from the sound pressure on a sphere

$$
\begin{aligned}
\frac{p_i + p_s}{p_0} &= \mathrm{TF}_{sphere} \\
&= \left(\frac{1}{ka}\right)^2 \sum_{q=0}^{\infty} \frac{j^{q+1} P_q(\cos\varphi)(2q+1)}{J_q'(ka) - jN_q'(ka)}.
\end{aligned} \tag{5.1}
$$

This transfer path TF_{sphere} is given by the incident pressure p_i, the scattered pressure p_s and the free field pressure p_0. They can be determined by the

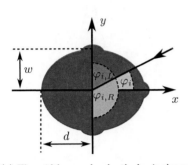

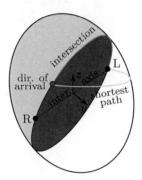

(a) The width w and a depth d_M is shown for the ellipsoid in the horizontal plane. The direction of the incident wave is measured by φ_i. The corresponding angles between φ_i and the ear are shown towards the interaural axis ($\varphi_{i,R}$ and $\varphi_{i,L}$).

(b) Considering the sliced ellipsoid for the incident wave (grey dot) at $(\theta, \varphi) = (20°, 285°)$, the path towards the averted left ear (black dot) on the intersection line is not the shortest possible path. Due to the depth of the head, which is smaller than the height, the shortest path is along the forehead.

Figure 5.1.: Ellipsoid model to derive an individualized ITD.

sum of the Bessel and Neumann functions (J'_q and N_q) as well as the Legendre polynomial P_q which are dependent on the direction, the sphere radius a and the wave number k. From the time difference of the direct sound at the right and left ear[1], the ITD can be calculated (see Section 5.2 for further details).

The sound pressure on a sphere is often based on rigid surface properties which can be reasonably assumed for the human skin (Katz, 2001). Nevertheless, small differences between the analytic model with rigid surfaces and measured HRIRs are to be expected.

The analytic solution is only one possibility to model the ITD (cf. Section 5.5). However, it is a very fast solution and therefore it is now used for the ellipsoid model. This model has already been published by Bomhardt and Fels (2014) and Bomhardt et al. (2016b).

The ITD is calculated by the ellipsoid model with a direction-dependent radius by reason of the fact that the head is more an ellipsoid with a width w, a depth d_M and a height h than a sphere. The geometric center of the ellipsoid is defined on the interaural axis between both ears[2]. From this center, the x-axis is positively defined towards the nose (natural viewing) as depicted in Fig. 5.1.

---

[1] The source positions on the sphere are chosen according to the position of the right and left ear of a human head.

[2] If necessary, it is also possible to consider an ear offset towards the back.

To derive the elevation angle-dependent radius, the ellipsoid has to be sliced by a plane. This plane is defined by three non-colinear points. Two of them are located at the ear canal entrances $y = \pm w$ and the third one is defined at the point on the ellipsoid where the incident wave arrives first.

The resulting slice of the plane and ellipsoid is an ellipse with an elevation-dependent depth $d_{ellipse}$ (see Fig. 5.1). On this ellipse, the radius

$$a_{ear}(\varphi_i) \quad = \quad \overline{r_{ellipse}(\varphi)} \quad \text{for} \quad \varphi \in [\varphi_i \ldots \varphi_{ear}] \tag{5.2}$$

is calculated for (5.1) by averaging the elevation angle-dependent radius

$$r_{ellipse}(\varphi) \quad = \quad \frac{w}{\sqrt{1 - \left(\frac{d^2_{ellipse} - w^2}{d^2_{ellipse}} \cdot \cos^2 \varphi\right)^2}} \tag{5.3}$$

of the ellipse between the incident direction φ_i and the ear position $\varphi_{i,ear} = \{\varphi_{i,L}, \varphi_{i,R}\}$ (Bronstein and Semendjajew, 1991, pp. 221-222).

Due to the fact that the intersection of the ellipsoid and the plane does not always describe the shortest path from the incident wave to the ear canal entrances, the ITD is larger than assumed for directions close to the interaural axis as in Fig. 5.1. This error can be reduced by interpolation of the radius a in the area close to the averted ear $\varphi_R = 70° \ldots 110°$ and $\varphi_L = 250° \ldots 290°$ to make sure that the radii from the analytic model match with the averaged radius of the shortest path (Bomhardt and Fels, 2014).

5.2. Interaural Time Delays

For the reason that the ITD cannot be measured directly, it has to be derived from HRIRs or HRTFs. Afterwards the calculated ITD can be used to adapt or reconstruct the ITD of an HRTF data set.

5.2.1. Estimation of the Interaural Time Delay

Different approaches to determine the ITD exist. To briefly recap, the most important approaches are presented: Phase delay (PD), interaural cross-correlation (IACC) and threshold method (THX). A comprehensive overview can be found in the work of Katz and Noisternig (2014).

Phase Delay The ITD_{PD} can be derived from the time of arrival (TOA) which is the delay τ of the arriving sound from a source at the ear. This delay can be determined from the unwrapped phase ϕ_{ear} of the HRTFs

$$\tau_{ear} = \frac{\phi_{ear}}{2\pi f} \tag{5.4}$$

dependent on the ear side $ear = \{R, L\}$. Subsequently, the ITD has to be calculated from the TOA τ at the right and left ears

$$\text{ITD}_{\text{PD}} = \tau_R - \tau_L. \tag{5.5}$$

Threshold One opportunity to derive the ITD in the time-domain is the detection of the onset of the HRIR. Often these onsets are not detected at their maximum but rather several decibels below this maximum. This makes the threshold level detection more robust since HRIRs close to the interaural axis have multiple peaks which are caused by the multi-path propagation to the averted ear.

According to (5.5), the ITD_{THX} is calculated from the time difference between the onset of the right and left ears.

Interaural Cross-Correlation Another possibility in time-domain is to calculate the ITD directly from the HRIRs. For that the interaural cross-correlation

$$\text{ITD}_{\text{IACC}} = \arg\max_{\tau} \left(\text{HRIR}_L \star \text{HRIR}_R \right). \tag{5.6}$$

is determined. This cross-correlation of HRIR_L and HRIR_R shows sometimes a maximum τ which accords to the time difference between both ears.

Comparison Compared to the determination of the ITD by the HRIRs, the ITD_{PD} is frequency-dependent since the phase of the HRTF is frequency-dependent. The ITD_{PD} is, however, prone to discontinuity of the unwrapped phase which leads to errors[3] (Bomhardt and Fels, 2014).

In contrast to this, the discontinuities of the ITD of the proposed time-domain approaches are often caused by discrete sampling. A smoother curve progression can be achieved by an interpolation of the time signals. Applying a low-pass filter also reduces rapid fluctuations of the HRIRs. The influence of such a low-pass filter is depicted in Fig. 5.2: For the non-filtered ITD_{IACC} more peaks are observable than for the filtered ones. Alternatively, the energy envelope of the

---

[3]The wrapped phase is a discontinuous function due to the fact that the phase is determined by the argument of a complex number which is defined either for interval $(-\pi, \pi]$ or $[0, 2\pi)$. These discontinuities can be removed by adding or subtracting 2π at these points. Nevertheless, fast increasing or decreasing phases may still show discontinuities (Tribolet, 1977).

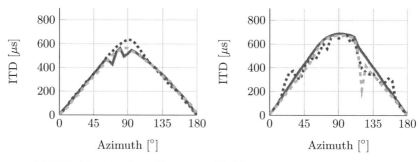

(a) ITD without bandpass filter.

(b) A low-pass Butterworth filter with a cut-off frequency of 2 kHz is applied.

Figure 5.2.: The ITD of subject # 17 is plotted for the phase delay (PD: solid line), the IACC (dashed line) and the threshold (THX: dotted line) against the azimuth angle.

HRIR helps to further diminish such fluctuations (Katz and Noisternig, 2014). The filtered ITD_{THX} shows more outliers than the filtered ITD_{IACC} due to the fact that the filtered HRIRs show no well-defined peak for the direct sound. Eventually, the ITD_{PD} and ITD_{IACC}[4] show smooth curve progressions for frequencies below 2 kHz. Due to the importance of the ITD for localization in the frequency range below 2 kHz (Middlebrooks and Green, 1991), the ITD_{PD} between 0.2 and 2 kHz is considered in the following.

Correction of misaligned Interaural Time Differences Despite careful alignment of each subject's head in the measurement setup (see Section 4.1), azimuth-offsets of the ITDs are observable in Fig. 5.3. If the right and left side of the head are not exactly symmetric, the incident waves from a sound source at $\varphi = 0°$ will not necessarily arrive at the same time at both ears. However, most of the heads are almost symmetric (see Section 7.3) therefore a delay between the incident waves of the right and left ear for a source at $\varphi = 0°$ is caused by a misalignment of the subject. For the measured HRTFs in the present database, the ITD offset of all subjects was $-17 \pm 13\,\mu s$.

If the position of the subject in the measurement setup has a translation error in x-direction, the ITD does not change much since the lengths of transfer paths towards the ears are comparable. On the other hand, a translation error of the subject's position in y-direction will lead to an ITD offset. Assuming a symmetric head, this offset can be monitored by unequal unsigned extrema of the ITD.

---

[4]In case the HRIRs are interpolated.

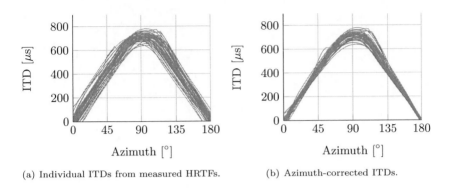

(a) Individual ITDs from measured HRTFs. (b) Azimuth-corrected ITDs.

Figure 5.3.: The original measured ITDs in the horizontal plane and the corrected ITDs are shown.

For an initial rotation error of the subject's position in azimuth, the ITD is not zero at $\varphi = 0°$ and $180°$ (Bomhardt et al., 2016b). In the regions of $\varphi = 0°$ and $180°$, the ITD can be approximated by a first order polynomial to avoid uncertainties. Subsequently, the deviation of the zero crossings of these approximated lines from $\varphi = 0°$ and $180°$ define the offset φ_{Off}. The averaged offset of all subjects was $\varphi_{Off} = -1° \pm 3°$. In the following investigations, the azimuth-error is corrected since this measure reduces the mismatch between measured and modeled ITD. However, the frontal and rear offsets differ slightly due to movements of the subject during the measurement.

5.2.2. Adaptation of Interaural Time Difference

Either TOA or ITD of an existing HRTF data set can be adapted by a fractional delay (Laakso et al., 1996) to minimize the mismatch between the given HRTF data set and the individual one of an arbitrary person. For both possibilities frequency-constant time delays[5] $\Delta\tau$ are assumed to adapt the ITD. Subsequently, the time delay $\Delta\tau$ is derived from frequency-averaged TOAs $\overline{\tau_{ear}}$ from a subject to be adapted (referred to as *target*) and an existing data set (referred to as *reference*)

$$\Delta\tau_{ear} = \overline{\tau_{ear,ref}} - \overline{\tau_{ear,target}}. \tag{5.7}$$

In general, the HRTF data set of the subject to be adapted is not given. Consequently, the optimal delay $\Delta\tau$ to adapt the given HRTF data set for this subject is

---

[5]Since the ITD_{PD} below 2 kHz does not show a strong frequency-dependency (see Kuhn (1977) for further details), the frequency-averaged delay τ is used.

not known. To adapt the given HRTF data set for this subject, an anthropometric ITD model can be used. Using the head dimensions of the existing HRTF and the one of the subject to be adapted, both ITDs can be estimated. Considering these estimated ITDs, the delay $\Delta\tau$ can be determined by

$$\Delta\tau_{ear} = \text{ITD}_{ref} - \text{ITD}_{target}. \tag{5.8}$$

In this case only the time delay $\Delta\tau$ of the averted ear has to be estimated to adapt the ITD[6].

If the ITD$_{\text{PD}}$ is used or the HRIRs are interpolated, the time delay τ_{ear} is in general an angle-dependent fractional delay. This delay can be realized using a simple FIR filter with a Lagrange interpolation (Laakso et al., 1996) which has a smooth magnitude response in the frequency-domain compared to other approaches. Nevertheless, the magnitude of the spectrum will decrease at higher frequencies which results in a tolerable[7] decline above 17 kHz.

The advantage of the adaption by a fractional shift $\Delta\tau_{ear}$ is that it maintains the fine structure of the ITD and the offset of the ears of the reference head. It is important to maintain this fine structure since the shoulder reflections improve the localization (Algazi et al., 2001a), and the human auditory system is sensitive to interaural differences of ongoing fluctuating envelopes in the time-domain above 3 kHz (McFadden and Pasanen, 1976). On the other hand, an individual adaption of delayed reflections as the shoulder reflection is not possible. Additionally, this method adapts an existing ITD and does not reconstruct the phase of an HRTF data set.

5.2.3. Reconstruction of Interaural Time Difference

If the phase of an existing HRTF data set has to be reconstructed completely by a minimum phase plus delay, the estimated TOA or ITD can be used as a time-constant delay τ_{ap}. As the delay of the minimum phase of the data set is generally not zero, the delay τ_{ap} has to be calculated from the ITD minus the delay of minimum phase

$$\tau_{ap} = \text{ITD} - \tau_{min}. \tag{5.9}$$

Otherwise, the resulting estimated ITD is over- or underestimated due to the minimum phase.

---

[6]Naturally, the ipsilateral HRIRs can be adapted as well.

[7]Provided that the subject has no hearing loss, the frequency limit of human hearing is around 20 kHz (Blauert, 1997, p. 2). Aggravating this situation, the measurement uncertainties due to the short wave lengths above 17 kHz have a strong impact.

5.3. Empiric Interaural Time Difference Model

Besides the analytic ellipsoid model, the possibility of deriving an anthropometric direction-dependent ITD description from the measured HRTFs is provided by the HRTF database presented (Bomhardt et al., 2016a). To be independent of the measurement grid, the ITDs are fitted with a polynomial[8]. It revealed that a polynomial of the fourth order

$$\hat{\text{ITD}}_{90} = \sum_{i=1}^{4} \alpha_i \cdot \varphi^i \quad \text{for} \quad 0° \leq \varphi < 90° \tag{5.10}$$

reduces the root mean squared error between the matched and original ITD sufficient for azimuth $0° \leq \varphi \leq 90°$ to $1.7\,\text{ms}$. For lower polynomials the error is larger (6.3 - $11.5\,\text{ms}$) especially close to the interaural axis.

Fitting the ITDs of all data sets of the database provides subject-dependent coefficients α. As the ITD depends on the head dimensions, a linear regression analysis was performed to express these coefficients α by the head width w and the depth d_M. The head height and ear offset are neglected since these dimensions did not improve the current fitted ITD. For an azimuth larger than $90°$, this function has to be adapted as follows

$$\hat{\text{ITD}}_{180}(\varphi) = \quad \hat{\text{ITD}}_{90}(180° - \varphi) \quad \text{for} \quad 90° \leq \varphi < 180°, \tag{5.11}$$

$$\hat{\text{ITD}}_{270}(\varphi) = \quad -\hat{\text{ITD}}_{90}(\varphi - 180°) \quad \text{for} \quad 180° \leq \varphi < 270°, \tag{5.12}$$

$$\hat{\text{ITD}}_{360}(\varphi) = \quad -\hat{\text{ITD}}_{90}(360° - \varphi) \quad \text{for} \quad 270° \leq \varphi < 360°. \tag{5.13}$$

To consider the elevation-dependency of the ITD, (5.10) - (5.13) are extended by

$$\hat{\text{ITD}}(\theta, \varphi) \quad = \quad \hat{\text{ITD}}(\varphi) \cdot \sin(\theta). \tag{5.14}$$

Due to the decreasing localization performance for upper and lower directions (cf. Section 2.1), it is assumed that the sinus-extension[9] is sufficient for directions out of the horizontal plane.

5.4. Evaluation Standards of Interaural Time Difference Models

The comparison between modeled and measured ITDs can be performed either analytically or subjectively. Meanwhile the analytic evaluation determines the error; the subjective evaluation shows whether this error is audible. For this reason, the current section describes on the one hand, an analytic procedure to

---

[8]The algorithm is based on the best fit in a least-squares sense.

[9]This sinus-extension is introduced by Savioja et al. (1999) (cf. (5.22)).

determine the ITD error of a whole HRTF data set and on the other hand it deals with a listening test which determines the just noticeable ITD error.

5.4.1. Analytic Evaluation Standards of Interaural Time Difference Models

To evaluate different ITD estimation models, the error between the measured and modeled ITD

$$\Delta \text{ITD} \quad = \quad \text{ITD}_{Meas} - \text{ITD}_{Model} \tag{5.15}$$

is one possibility, but also the correlation coefficient between the ITD maxima can be investigated.

Since the ITD is direction-dependent, it is difficult to compare different models for every single direction. Therefore, the mean ITD error and its standard deviation can be expressed as a weighted sum of errors

$$\overline{\Delta \text{ITD}_{\theta,\varphi}} \quad = \quad \frac{\sum\limits_{k=1}^{K-1} \sum\limits_{l=1}^{L} \beta_{k,l} \cdot \Delta \text{ITD}_{k,l}}{\sum\limits_{k=1}^{K-1} \sum\limits_{l=1}^{L} \beta_{k,l}} \tag{5.16}$$

where k and l are the points (Bomhardt et al., 2016b)[10] on the measurement sphere and β are the surface weights (Leishman et al., 2006). Subsequently, the standard deviation of the overall error is defined as

$$\sigma_{\Delta \text{ITD}_{\theta,\varphi}} \quad = \quad \sqrt{\frac{\sum\limits_{k=1}^{K-1} \sum\limits_{l=1}^{L} \beta_{k,l} \cdot \left[\Delta \text{ITD}_{k,l} - \overline{\Delta \text{ITD}_{\theta,\varphi}} \right]^2}{\sum\limits_{k=1}^{K-1} \sum\limits_{l=1}^{L} \beta_{k,l}}}. \tag{5.17}$$

5.4.2. Subjective Evaluation Standards of Interaural Time Difference Models

Dependent on the study, the just noticeable ITD error is $16\,\mu s$ for frontal directions (Aussal et al., 2012) and up to $125\,\mu s$ for lateral directions (Simon et al., 2016). Due to these large angle-dependent deviations, the error was investigated in a listening experiment for the subjective evaluation of the ITD models. The following experiment design was inspired by the study of Simon et al. (2016)

---

[10]The lowest measurement directions are weighted stronger than all others because of the gap in the lower sphere in the measurements of the database (Bomhardt et al., 2016a). For this reason, these directions are ignored for the mean and standard deviation.

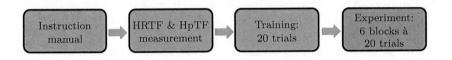

Figure 5.4.: The listening experiment consisted of four blocks.

who used different types of alternative forced choice tests to determine the just noticeable ITD mismatch.

Just noticeable ITD error In total, 32 subjects, who were on average 25 ± 4 years old, were tested. Most of them (27) were right-handed, 23 had no previous experience with binaural reproduction techniques and half of them were male. All of the subjects reported normalhearing.

The listening experiment was split into four parts (see Fig. 5.4): Reading the instructions, the measurement of the individual HRTFs and HpTFs, a practice run and the main experiment.

The transfer functions were measured according to the procedure described in Sections 2.6 and 4.1. The measurements and experiment took place in a low-reflection room. Six loudspeakers (Genelec 6010A) in the horizontal plane from 270° to 345° and Sennheiser KE3 microphones, which were supported by a dome at the ear canal entrance, were used to measure the HRTFs. To recap, these directions were chosen because of the right ear advantage (cf. Section 2.1). The distance between the loudspeakers and the center of the subject's head was 1.5 m. According to measured HRTF of the database (Bomhardt et al., 2016a), the measured transfer function was time-windowed and cropped to 256 samples. Afterwards, the HpTFs were measured with KE3 microphones and headphones HD650 by Sennheiser. For this purpose, the subject had to reposition the headphones eight times on the head. An averaged HpTF was calculated using the inverted mean plus twice the standard deviation of these measurements (Masiero and Fels, 2011).

Since the localization performance in the rear hemisphere is comparable to the frontal hemisphere, only the frontal directions were tested (cf. Section 2.1). Furthermore, localization performance decreases out of the horizontal plane. Subsequently, it was assumed that the noticeable ITD error will increase towards these directions.

The stimuli were generated using pink noise with three pulses with a length of 0.3 s and a pause of 0.1 s which were convolved with the measured HRIRs and headphone impulse responses. The main experiment consisted of six rounds

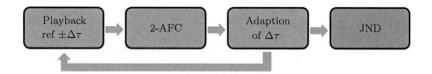

Figure 5.5.: Each block of the listening experiment consisted of the first three steps to determine the JND of the ITD. The first three steps were repeated 20 times in each block.

respectively to the measured HRTF directions. In each round the ITD was manipulated: Either a delay was added or subtracted. If the delay is subtracted and the difference is noticeable, the subject is able to identify that the sound is shifted towards the front. If the delay is added, it will be shifted towards the interaural axis. So, the task for the subject was to identify whether the stimuli was shifted to the front or to the side compared to the stimuli convolved with the original HRTF.

The procedure in each experiment block is illustrated in Fig. 5.5: In the first trial of each block the subject heard the reference and the manipulated stimulus with an initial delay of $100\,\mu s$. Then the subject had to decide by a two-alternative forced choice (2-AFC) whether the second stimulus was more to the front or back[11]. Depending on whether the answer was right or wrong, the delay is adapted by the QUEST method (Watson and Pelli, 1983). A correct answer resulted in a shorter delay in the next trail and a wrong answer in a longer delay. The QUEST method is an adaptive psychometric procedure which starts with an initial probability distribution (see Fig. 5.6). The mean of this distribution was $100\,\mu s$ and its standard deviation $100\,\mu s$ to cover large variations. This distribution was multiplied by the psychometric function which took the guessing probability $\delta = 0.1$, the false alarm rate $\gamma = 0.5$ and additionally a slope $\beta = 3$ into account. Furthermore, this psychometric function depended on the answer. If the answer was correct, it was the mirrored version of the false one (see Fig. 5.6). Finally, the resulting mean distribution was shifted to lower or higher test values. In this experiment the mean value of the current distribution was used to determine the next delay. After 20 trials the tested delays varied only slightly and can be assumed as converged. Therefore, the last tested delay was the desired just noticeable ITD error.

The adaption by the QUEST method has the advantage that it is faster and more

---

[11]The direction of the shift was chosen randomly.

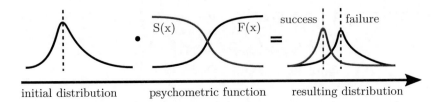

initial distribution   psychometric function   resulting distribution

Figure 5.6.: The adaptive QUEST method is a psychometric procedure to determine subjective threshold levels. It calculates the testing threshold from a probability distribution and adapts this distribution using the psychometric function based on the answers of the subject.

precise than testing a set of discrete thresholds. The disadvantage is that the psychometric function and initial distribution have to be estimated beforehand. This was done by pretests.

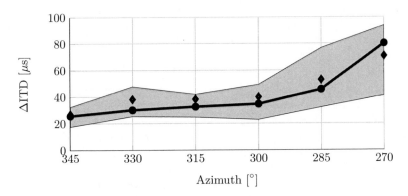

Figure 5.7.: The just noticeable ITD error is plotted direction-dependently by the median of all subjects (black line). The grey area marks the interquartile range. The subject-averaged just noticeable ITD error is marked by a diamond.

The results of 24 subjects, who performed all test conditions[12], are depicted in Fig. 5.12. The JND is almost stable between $300°$ and $345°$ and grows faster for lateral angles. In the present experiment no significant gender effects are detected. Only a slight advantage of females is observed.

The study by Simon et al. (2016) investigated the JND of the ITD in the horizontal

---

[12]The remaining eight subjects were tested six times at $\varphi = 345°$.

plane at $30°$ and $90°$ for different protocols. Without going into precise details of these protocols, the median of the JND at $30°$ varies between 24 and $44\,\mu s$ and the JND at $90°$ between 68 and $125\,\mu s$ dependent on the protocol. The median of the JND at $30°$ of the present study is $30\,\mu s$ and at $90°$ it amounts to $80\,\mu s$ (deviations from this median are shown in Fig. 5.7). Both JNDs are in line with those of the study by Simon et al. (2016).

5.5. Review of Interaural Time Difference Models

Analytic Solutions One of the first well-known models by Woodworth (1940) is related to the sound pressure on the surface of a rigid sphere

$$\text{ITD}_{\text{Wood}} = \frac{a}{c_0} \left(\sin \varphi + \varphi \right) \quad \text{for} \quad 0 \leq \varphi \leq \frac{\pi}{2} \tag{5.18}$$

for low frequencies $1 < ka < 5$. If a mean head radius of $a = 87.5\,\text{mm}$ is assumed, (5.18) is valid from $0.6\,\text{kHz}$ up to $3.1\,\text{kHz}$. Below $(ka)^2 \ll 1$ the following equation (Kuhn, 1977) is more precise

$$\text{ITD}_{\text{Kuhn}} = \frac{3a}{c_0} \sin \varphi. \tag{5.19}$$

Comparing human and spherical ITDs, the spherical ITD is often smaller than the measured human ITD due to the fact that the volume of the sphere is smaller than that of the human head (Katz, 2001). The relation between the ITD and the head parameters showed that a human ITD can be better estimated using an optimal radius

$$a_{\text{Algazi}} = 0.51w + 0.18d + 0.019h + 32\text{mm} \tag{5.20}$$

based on the CIPIC database (Algazi et al., 2001c). This optimal radius shows a very small influence of the head height h and an additional bias of $32\,\text{mm}$ (Algazi et al., 2001c; Busson, 2006). Another possibility to find an optimal radius is a subjective real time adjustment of $\text{ITD}_{\text{Woodworth}}$ (Lindau et al., 2010). Subsequently, the scaling factor between the human and spherical ITD can be derived from the results of the test with regard to the head width w.

Since the ITD is elevation- and azimuth-dependent, (5.18) and (5.19) can be extended

$$\text{ITD}_{\text{Larcher}} = \frac{a}{c_0} \left(\arcsin \left(\sin \varphi \cos \theta \right) + \sin \varphi \cos \theta \right). \tag{5.21}$$

to consider that the ITD becomes smaller for upper and lower directions (Larcher and Jot, 1997). This formula can be further simplified to

$$\text{ITD}_{\text{Savioja}} = \frac{a}{c_0} \left(\sin\varphi + \varphi \right) \cos\theta \qquad (5.22)$$

as described in a study by Savioja et al. (1999).

The position of the human's ears and the diffraction on the head cannot be considered completely by a spherical approach. For this purpose, the delay between both ears can also be estimated by a geometric model of the head with an ear offset (Busson, 2006; Ziegelwanger and Majdak, 2014).

Empiric Solutions Empiric approaches such as spatial Fourier analysis in combination with multiple regression by anthropometric dimensions can be used to cover front-back or right-left asymmetries (Zhong and Xie, 2007). Another option is either the decomposition of the ITD into principal components[13] (Aussal et al., 2012) or spherical harmonics (Zhong and Xie, 2013a) to take these asymmetries into account and estimate the ITD from head and pinna dimensions.

Numeric Solutions Numeric calculations provide another opportunity to consider asymmetries of the head. The front-back asymmetry can be taken into account by the width, the depth, the height and the ear offset of a numeric ellipsoid model (Duda and Algazi, 1999).

If the ITD is determined in the time-domain (cf. Section 5.2), this ITD is mainly influenced by the first arriving wave front TOA. Consequently, the ITD does not consider the delayed shoulder reflections. These reflections arrive delayed and damped due to the detour via the shoulder to the ear. In the frequency-domain, the shoulder reflection can be observed as a destructive interference between 1 kHz and 2 kHz. This minimum is especially important for the localization at the contra lateral ear (Algazi et al., 2001b). Numeric models such as the snow man model (Algazi et al., 2002a,b) with a head and torso take these reflections into consideration.

Presented Interaural Time Difference Models The ellipsoid model presented provides an analytic solution which depends on the head height, depth and width. Consequently, it does not support ear offsets towards the back or shoulder reflections. However, if this model is used to adapt the ITD of an existing HRTF data set as proposed in Section 5.2.1, the properties as ear offset and shoulder

---

[13]The principal component analysis is applied on the complex-valued spectrum in Chapter 7. The resulting complex-valued components and their corresponding scores can be used to reconstruct or estimate the ITD (see Section 7.5.1 for further details).

reflections are maintained but cannot be individually adapted.

The same applies to the proposed empiric ITD model in Section 5.3, which does not support any head asymmetries.

5.6. Comparison of Interaural Time Difference Models

In this section, the comparison of the presented models and the following models is carried out in four steps:

1. The difference of the measured and modeled ITDs in the horizontal plane is discussed first. For this purpose, the general curve progressions as well as the maximum ITDs are observed.

2. The error of the measured and modeled ITDs is discussed angle-dependently.

3. The overall mismatch of measured and modeled ITDs is discussed using (5.16) and (5.17).

4. The analytic ITD error is evaluated using the just noticeable ITD error from Section 5.4.2.

5.6.1. Analysis of Interaural Time Difference Models in the Horizontal Plane

For the comparison the analytic models are considered:

$ITD_{Ellipsoid}$: The radius a_{ear}, which depends on the incident direction, the width, depth and height, is used to calculate the HRTF of an ellipsoid by (5.1). Subsequently, the ITD_{PD} of this ellipsoid is estimated by the phase.

ITD_{Adapt}: A randomly chosen HRTF data set (#17) is adjusted using the ellipsoid model. For this purpose, the $ITD_{Ellipsoid}$ is calculated once from the anthropometric dimensions of subject #17 and it is additionally determined from the anthropometric dimensions of the subject to be adapted. The difference of both $ITD_{Ellipsoid}$s is used to adapt the HRTF data set #17 (cf. Section 5.2.1).

$I\hat{T}D$: The empiric $I\hat{T}D$ (5.13) is estimated from the anthropometric dimensions of the subjects.

ITD_{Wood}: The Woodworth approach (5.18), which is valid up to 3.1 kHz, is used to estimate the ITD by the mean head radius of the subjects.

ITD_{Kuhn}:   Kuhn's approach (5.19), which is valid below 0.6 kHz, is used to estimate the ITD by the mean head radius of the subjects.

$\text{ITD}_{\text{Kuhn},opt}$:   The optimal radius $\hat{a} = (0.76w + 0.31d_M)/1000$ is determined from the database (Bomhardt et al., 2016a) by a linear regression analysis (see Appendix B)[14]. Consequently, this optimized radius is used to calculate $\text{ITD}_{\text{Kuhn},opt}$.

The curve progressions of the measured and modeled ITDs of 47 subjects in the horizontal plane will give a first impression of the differences between both (see Fig. 5.8). The ITD of the measured and adapted HRTFs is thereby derived from the phase of the HRTF. To obtain a time-constant delay (5.4), the TOA is averaged between 0.2 and 2 kHz due to the importance of the low frequent ITD (Macpherson and Middlebrooks, 2002) and the almost frequency-independent ITD in this range (Kuhn, 1977).

In the upper plots of Fig. 5.8 it can be observed that the ellipsoidal ITD is smaller than the measured ITD. Additionally, the maximum seems to be wider so that a more detailed study of this mismatch has to be done in Section 5.6.2. Using the ellipsoid to adapt the ITD of an HRTF, the general deviations of the curve progressions are no longer identifiable in Fig. 5.8[15]. The fitted ITDs are comparable to the measured ones in Fig. 5.8. Nevertheless, the width of the fitted ITD maxima of small heads seems to be smaller than the measured ones. Additionally, the curve progression is symmetric to $90°$ due to the neglected ear offset. The same applies to the models of Woodworth and Kuhn. The Woodworth model, which is valid for frequencies $f > 600\,\text{Hz}$, underestimates the measured frequency-averaged ITD. The maxima of ITD_{Kuhn} seem to be similar to the measured ones. The width of the maxima of Woodworth's model is smaller than the ones of the measured ITDs. In contrast to this, Kuhn's estimation shows a wider maximum than the measured ITDs. These differences for (5.18) and (5.19) can be traced back to the phase delay method which calculates a frequency-averaged ITD in the range between 0.2 and 2 kHz.

The main inter-subject variations of the ITD can be found at its maximum. These variations are mainly forced by head dimensions. To evaluate the precision of the modeled ITD maximum, it is plotted in relation to measured ITD in Fig. 5.9. There and in the Table 5.2, it can be observed that the ellipsoid underestimates the ITD. This underestimation of 5 to 10% is already displayed in other previous

---

[14]The optimized radius of Algazi (5.20) does not match the current head dimensions because of different measurement points. The head height and offset are neglected owing to the low influence $0.76w + 0.31d \approx 0.69w + 0.31d_M + 0.03d_U + 0.04h$.

[15]Although no difference is observable in Fig. 5.8, the horizontal error (5.15) of the models in Fig. 5.11 shows that a mismatch between the adapted and individual HRTFs remains.

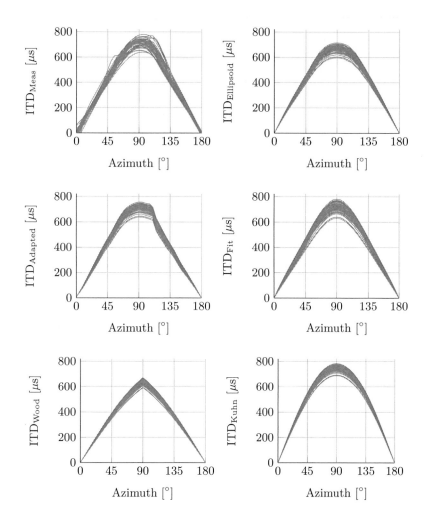

Figure 5.8.: The measured ITD and estimated ITD are compared in the six plots in the horizontal plane for 47 subjects.

studies (Duda and Algazi, 1999). Adapting the ITD of an existing HRTF data set does not show this underestimation. Moreover, the adapted ITD_{Adapt} indicates a linear relation between the measured and adapted ITD since the relative error is almost 0% and the slope m of the fitted curve (see Table 5.2) is close to one. Using (5.20), the ITD_{Wood} underestimates the measured ITD whereas the maximum of ITD_{Kuhn} fits well with those of the measured ITDs. The correlation between all modeled and measured maximum ITDs is almost similar and shows that the maximum ITD can be well modeled by the head width and depth. To summarize the mismatch of the models, the mean error $\mu = \overline{|\max ITD_{Meas} - \max ITD_{Model}|}$ between the maxima are regarded in Table 5.2. For the adapted ITD, the fitted ITD and Kuhn's model with an optimized head radius, the mean error is below $15\,\mu s$. Only the ellipsoid and Woodworth's approach show larger deviations. The standard deviation is similar for all models with approximately $11\,\mu s$.

Model	μ [μs]	σ [μs]	Rel. error [%]	$m \cdot ITD_{Meas} + n$
Ellipsoid	75	13	10	0.9 ITD $+$ 32 μs
Adapted	10	9	1	0.9 ITD $+$ 92 μs
Fit	15	12	2	1.1 ITD $-$ 58 μs
Woodworth	94	14	13	0.6 ITD $+$ 200 μs
Kuhn	15	10	-2	0.7 ITD $+$ 233 μs
Kuhn opt.	11	10	0	1.1 ITD $-$ 80 μs

Table 5.2.: The mean error $\mu = \overline{|\max ITD_{Meas} - \max ITD_{Model}|}$ and its standard deviation σ is shown in the second and third columns. The fitted curve describes the linear relationship between the modeled ITD and the measured ITD.

Comparing the ITDs of one subject for repeated HRTF measurements[16] shows for subject #1 a mean ITD maximum of $710 \pm 5\,\mu s$ and for subject #17 a mean ITD maximum of $674 \pm 4\,\mu s$. Hence, the standard deviation of the ITD of the repeated measurements is very small compared to the inter-subject deviations (cf. Fig. 5.9) or the ITD mismatch (cf. Table 5.2).

5.6.2. Angle-Dependent Analysis of Interaural Time Difference Models

Since the ITD is not only restricted to the horizontal plane, the error between measured and modeled ITDs are discussed direction-dependently in Fig. 5.10. Due to the fact that the Woodworth and Kuhn models can be extended by an

---

[16]The HRTFs of subjects #1 and #17 were measured four times on different days with a reconstructed measurement setup.

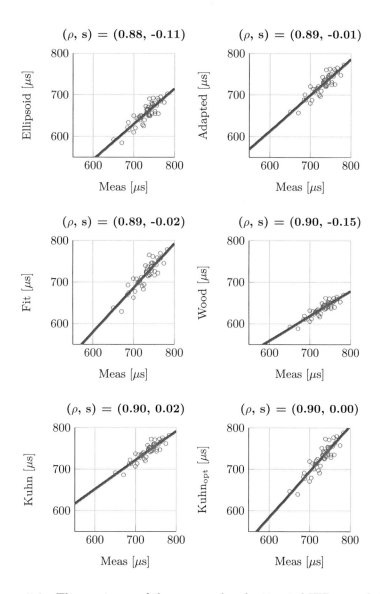

Figure 5.9.: The maximum of the measured and estimated ITDs are plotted. The solid line is a linearization which can be found in Table 5.2. Additionally, the correlation coefficient ρ and a scaling factor s between measured and modeled ITD is given in the title.

elevation-dependent term as in (5.21) or (5.22), these two models are discussed in the following. Additionally, the adapted ITD from the ellipsoid model as well as the fitted ITD are considered in Fig. 5.10.

In general, an angle-dependent mean error (5.15) over all subjects between the models and the measurements is in line with ITD differences detected in the horizontal plane (cf. previous Section 5.6.1. The error of the adapted ITD is between -25 and $25\,\mu s$. In this case, the $\text{ITD}_{\text{Adapt}}$ slightly overestimates the averaged measured ITD in the lower rear hemisphere and underestimates it in the upper frontal hemisphere. This difference can be traced back to the torso and head shape which cannot be adapted by the ellipsoid model.

The error of the fitted ITD is slightly larger than the one of the $\text{ITD}_{\text{Adapt}}$ and is in the range between -50 and $50\,\mu s$. Small overruns above $\theta > 60°$ can be attributed to measurement errors and phase delay method uncertainties. In contrast to the adapted ITD, this model overestimates the ITD in the lower hemisphere and underestimates the ITD in the upper hemisphere. The error occurs asymmetrically to the coronal plane, indicating a missing ear offset.

In general, the models of Savioja et al. (1999) and Larcher and Jot (1997) underestimate the maximum. Differences occur especially close to the interaural axis. The Savioja model shows larger mismatches of the curve progression around $\varphi = 90°$ and $120°$. This error is also visible for the Larcher & Jot model but not that pronounced. The error of the Larcher & Jot model is generally larger than the one of the Savioja model. A slight asymmetry towards the coronal plane of the error is visible for both models.

Considering an optimized radius $\hat{a}$ for the Larcher & Jot and Savioja models minimizes the error but still shows the same asymmetries and deviations around azimuth angles $90°$ and $120°$.

5.6.3. Mean Angular Error of the Interaural Time Difference Models

Since the error between measurement and models has been discussed direction-dependently in the previous Section 5.6.2, it should now be regarded as an overall error by (5.16) and (5.17) in Table 5.3. The error is evaluated as a signed and unsigned error to avoid the averaging-effect (signed) but also shows the overall error tendency (unsigned).

In comparison, the adapted $\text{ITD}_{\text{Adapt}}$ has the lowest mean error and standard deviation due to the fact that the ITD of a measured HRTF data set is adapted. An ITD of a measured HRTF data set already implies a torso, ear offset and head asymmetries. Consequently, only a very slight overestimation of the averaged ITD can be observed.

In case where the ITD has to be recovered, the fitted ITD shows the most accurate

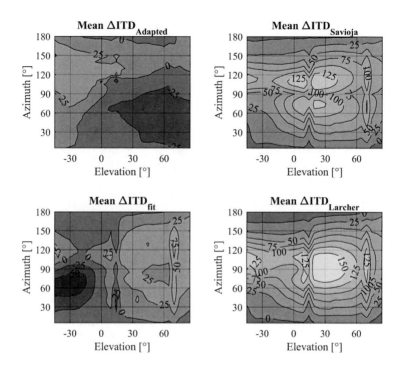

Figure 5.10.: The mean difference ΔITD between measured and modeled ITD is compared in the four plots direction-dependently over all 47 subjects. The error ΔITD is displayed in microseconds.

Model	Signed error		Unsigned error	
	$\overline{\Delta\text{ITD}}_{\theta,\varphi}$ $[\mu s]$	$\sigma_{\Delta\text{ITD}_{\theta,\varphi}}$ $[\mu s]$	$\overline{\Delta\text{ITD}}_{\theta,\varphi}$ $[\mu s]$	$\sigma_{\Delta\text{ITD}_{\theta,\varphi}}$ $[\mu s]$
Ellipsoid	31	47	48	33
Adapted	-5	24	29	11
Fit	10	27	31	12
Savioja	57	41	61	36
Kuhn $\hat{a}$	-24	39	43	24
Larcher	64	51	68	46

Table 5.3.: Comparison of the mismatch of the ITDs by (5.16) and (5.17) averaged over all 47 subjects. The error in the second column is considered signed and in the third column as unsigned.

solution. The error is slightly higher than for the adapted ITD.

Using Savioja's model (5.22) or Larcher & Jot's model (5.21) results in larger ITD errors than all other presented models. This applies for both the signed and unsigned error in Table 5.3.

Due to the fact that (5.19) showed good results in the horizontal plane for an optimized head radius, it is here extended

$$\mathrm{ITD}_{\mathrm{Kuhn},opt} = \frac{3\hat{a}}{c_0} \sin\varphi\cos\theta \qquad (5.23)$$

by the elevation-dependent part of (5.22). This solution shows a smaller error than the formula of Savioja and a similar unsigned error as in the ellipsoid. But the error is still larger than for the fitted ITD.

5.6.4. Subjective Evaluation of the Interaural Time Difference Models

Taking into account the results of the listening experiment in Section 5.4.2 and the results of Table 5.2 around the interaural axis makes it possible to distinguish whether an ITD mismatch is noticeable or not. The threshold[17] which is determined by the experiment is plotted as an area in Fig. 5.11. If the error between the modeled and measured ITD exceeds this area, it is audible.

A large noticeable underestimation of the ITD is only observed by Woodworth's model in Fig. 5.11, but also the ellipsoid model slightly underestimates the measured ITDs (see Fig. 5.11). In comparison to the underestimation of Woodworth's model, the one of the ellipsoid is in the tolerable range between 41 and 94 μs of the listening experiment. Interestingly, the maximum error of $\mathrm{ITD}_{\mathrm{Ellipsoid}}$ is smaller than in Table 5.2. This can be attributed to the fact that the maximum of $\mathrm{ITD}_{\mathrm{Ellipsoid}}$ is always at $\varphi = 90°$ whilst the measured one varies around $\varphi = 90°$.

Kuhn's model overestimates the measured ITD and exceeds the perception threshold especially in the ranges between 30° and 90° as well as 120° and 150°. Moreover, there is no significant difference if the head radius a_{Algazi} or $\hat{a}$ is used. The best results are achieved by the adapted and the fitted ITD. Their error is almost below the perception threshold.

In general, the underestimation of the ITD is not as critical as the overestimation. If the ITD is overestimated, the ITD could be larger than the maximum natural ITD and leads to a diffuse source, which is difficult to localize (Shinn-Cunningham et al., 1998).

---

[17] A front-back as well as a right-left symmetry is assumed for this threshold.

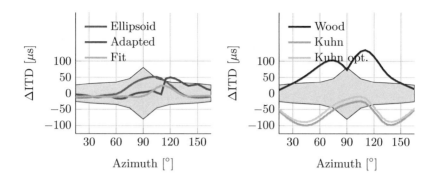

Figure 5.11.: The not audible ITD error from Fig. 5.7 marked by an area. Additionally, the mean error (5.15) is plotted for the considered approaches.

5.7. Influence of the Anthropometric Measurement Error on the Interaural Time Difference

The individualization of an HRTF data set by the models presented is only feasible if the anthropometric dimensions are known. Measurement devices such as measurement tape, caliper or scanner can be used. Nevertheless, this measurement equipment is prone to errors and will influence the accuracy of the adapted ITD.

Using the determined noticeable ITD error in combination with an ITD model shows how accurate the anthropometric input data has to be[18]. For this purpose, the dimension deviations of the head width Δw and depth Δd are discussed for (5.23). The formula of the fitted ITD, which is more accurate, is not taken into account due to the higher order derivatives (cf. (5.24)). Furthermore, only the uncertainties of the anthropometric dimensions are regarded at this point. The uncertainties of the positioning during the HRTF measurement are discussed in Section 5.2.2 and reduced by the analysis of the zero crossings and the symmetry of the ITD. The impact of movements of the subject during the measurement are small since the standard deviation of the maximum ITD of repeated measurements (two subjects and four repetitions per subject) amounted to $5\,\mu s$. According to the propagation of uncertainty of independent variables, the mea-

---

[18]This can be assumed if the ITD model fits the individual ITD accurately and the angular error can be neglected.

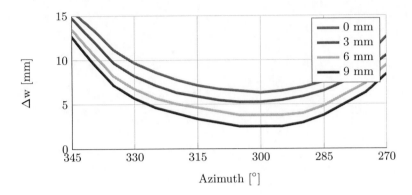

Figure 5.12.: To find the minimal allowed measured error for the head width and depth, the error ITD_r subtracted from the perception threshold $\Delta\text{ITD}_{median}$. The absolute value of $|\Delta\text{ITD}_{median} - \text{ITD}_r|$ is shown azimuth-dependently in the horizontal plane for a varied Δd.

surement error for the head width and depth

$$
\begin{aligned}
\Delta\text{ITD}_r &= \frac{\partial\text{ITD}_{\text{Kuhn},opt}}{\partial w} \cdot \Delta w + \frac{\partial\text{ITD}_{\text{Kuhn},opt}}{\partial d} \cdot \Delta d \\
&= \frac{3}{1000\, c_0} \left(0.75\Delta w + 0.31\Delta d\right) \sin\varphi \cos\theta \quad (5.24)
\end{aligned}
$$

can be compared with the median of the noticeable ITD error $\Delta\text{ITD}_{median}$ from the results of the listening experiment in Fig 5.12. The absolute value of the difference $|\Delta\text{ITD}_r - \Delta\text{ITD}_{median}|$ is taken to find the smallest allowed deviation

$$
\arg\min_{\Delta w} |\Delta\text{ITD}_r - \Delta\text{ITD}_{median}| \quad (5.25)
$$

for Δw or respectively for Δd. Regarding Fig. 5.12, the allowed error of Δw decreases with an increasing error Δd almost linearly. Therefore, both errors can be expressed as a linear combination

$$
\Delta w = -0.43\Delta d + 6.4\text{mm}. \quad (5.26)
$$

Assuming that $\Delta w \approx \Delta d$ will result in an allowed measurement error of $4\,\text{mm}$.

6

Interaural Level Difference

An ILD can be determined direction- and frequency-dependently from the HRTFs of the right and left and left ear

$$\text{ILD} = 20 \log_{10} \left(\frac{\text{HRTF}_L}{\text{HRTF}_R} \right).$$ (6.1)

This level difference between the ears is mainly caused by the head itself which shadows the averted ear. The smaller the wave lengths, the larger the ILD due to the shadowing effect. However, at frequencies larger than 2 kHz (Kulkarni et al., 1999; Shaw and Teranishi, 1968), the shadowing effect is not the only effect which influences the ILD. Creeping waves around the head as well as the pinna will impact the ILD as well.

In this chapter, the frequency- and direction-dependent behavior of an arbitrary human ILD is discussed as an example. Eventually, the influence of the head shape on the ILD is investigated for different subjects. Finally, a brief outlook is given how the ILD can be modeled by a geometrical approach.

Correction of misaligned Interaural Level Differences As previously mentioned in Sections 4.1 and 5.2.1, the misalignment of the subject in the measurement setup could lead to asymmetries of the ITD and ILD. Since the rotational shifts around the z-axis can be corrected by analyzing the ITD (see Section 5.2.1), the displacement can be corrected from the symmetry of the ITD or ILD.

Before the translative error was corrected for the present database (Bomhardt and Fels, 2017), the azimuth offset of the measured HRTF directions was adjusted according to the procedure described in Section 5.2.1. Afterwards, the ILD (6.1) was calculated frequency-dependently for horizontal directions. At frequencies below 2 kHz (Rayleigh, 1907; Kuhn, 1977; Kulkarni et al., 1999), the wave lengths are larger than the head for which reason the dimensions of the head have the most influence on the HRTFs. Therefore, the head itself can be assumed to be almost symmetric at frequencies below 2 kHz. Consequently, the ILD in the horizontal plane should be symmetric too. This symmetry was used to correct the level offset of the HRTFs by calculating the mean of the ILD in the horizontal plane. Subtracting this mean led to an almost symmetric ILD around the azimuth angle as for example in Fig. 6.1. The mismatch amounted to a subject-averaged level offset of $\Delta L = 1 \pm 4$ dB. The correction of the asymmetry enhanced the following analysis of analytic, numeric and measured ILDs.

6.1. Characteristics of the Human Interaural Level Difference

To describe the characteristics of the human ILD, the ILD of the arbitrarily chosen subject #17 of the present database (Bomhardt et al., 2016a) is calculated according to (6.1) and plotted in Fig. 6.1. Since this ILD is frequency- and direction-dependent, it is shown in Fig. 6.1 in the horizontal plane at 2.5 kHz. This frequency is chosen for the following discussion since humans use ILD cues above 2 kHz for the localization (cf. Section 2.1.1) but also because of the low impact of the pinna at this frequency (Shaw and Teranishi, 1968; Kulkarni et al., 1999). Later in this section, it will be shown that the pinna causes peaks and notches at higher frequencies which complicates the initial discussion of the ILD. Anyhow, the curve progression of the human ILD at 2.5 kHz in the horizontal plane is caused by the sound pressure at the ipsi- and contralateral ear. As depicted in Fig. 6.1, the sound pressure at the contralateral ear is lower than the pressure at the ipsilateral. Meanwhile the right ear is the contralateral one between 0° to 180°, the left ear becomes the contralateral one between 180° and 360°. Consequently, if the sound pressure of the left is subtracted from the one of the right ear, the resulting ILD is negative in the range from 0° to 180° and positive from 180° to 360°. The curve progression of this ILD has two characteristic notches close to the interaural axis which are mainly caused by the creeping waves around the head. These creeping waves may also be described as the diffracted wave around the shadowed surface. For human ILDs, the rear extrema between 90° and 270° are often larger than the frontal ones (see Fig. 6.2). This effect can be attributed to the ear offset towards the back of the head but also to the directivity of the ear itself (Lins et al., 2016; Bomhardt and Fels, 2016). Due to the fact that in the range of 180° to 360°, the left ear is the contralateral ear, the ILD features two maxima which are almost symmetric to the notches. Considering the initial HRTFs of the right and left ear in Fig. 6.1, it can be inferred that the notches and peaks were generated by the notches of the averted HRTFs. Compared to the contralateral HRTFs, the ipsilateral ones show a smooth increasing curve progression towards the interaural axis.

As mentioned before, the peaks and notches of the ILD increase at higher frequencies due to the shadowing of the head (see Fig. 6.2). This effect is also responsible for the side lobes above 3 kHz. Moreover, the directivity of the pinna and the first resonance of the cavum concha reinforce the extrema of the ILD.

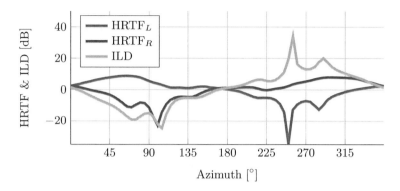

Figure 6.1.: The HRTFs of the right and left ear (HRTF$_R$ and HRTF$_L$) are plotted at 2.5 kHz for subject #17 against the azimuth angle. The ILD is calculated according to (6.1).

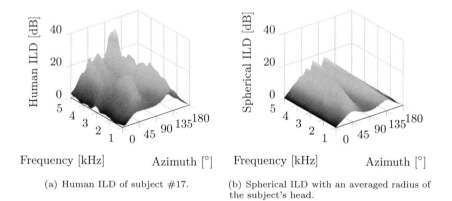

(a) Human ILD of subject #17.

(b) Spherical ILD with an averaged radius of the subject's head.

Figure 6.2.: The absolute value of human and spherical ILD is plotted against frequency and azimuth.

The difference between a head and a sphere phenomenon can be depicted in Fig. 6.2 where both are plotted dependent on the azimuth angle and the frequency. At low frequencies the ILD of both is very similar while at higher frequencies the maxima increase more intensely for the human than for the sphere. A detailed analysis of the difference between the human and spherical ILD will be given in the next section.

6.2. Influencing Anthropometric Dimensions

To study the inter-subject differences of human ILDs, the first 23 subjects of the database (Bomhardt et al., 2016a) are investigated in the following[1]. The influence of the head dimensions (ear offset, width and depth) and pinna is studied by simulations. Based on the head dimensions and the three-dimensional pinna models, the head was simulated as a sliced sphere with pinna. Eventually, the head was modeled as a sliced sphere without ears, and finally the head was approximated by a sphere (Bomhardt and Fels, 2016).

Determination of Analytic ILDs   Each head was approximated by a sphere with an averaged radius from the head's width and depth by (5.3). According to the reciprocity between source and receiver (Fahy, 1995), the sources were placed at the positions of the ear canal entrance (Fels et al., 2004). The receiver points were located on a 1-meter circle around the origin in the horizontal plane (see Fig. 6.3). Due to the fact that the origin on the interaural axis was not the geometric center of the head, the receiver points were shifted towards the back of the head[2]. Taking this setup with averaged radius and ear offset into account, the spherical ILD was calculated by (5.1) and (6.1) (Mechel, 2008, pp. 185-285). Due to the complex behavior of the ILD at higher frequencies, the current spherical ILDs are discussed at 2.5 kHz in the horizontal plane in Fig. 6.4. The only difference between the current ILDs and the spherical ILD in Fig. 6.2 is that the current ILD depends on the head radius and the ear offset of the 23 subjects. These subject-dependent spherical ILDs show two minima between $0°$ and $180°$ whose positions vary as a consequence of the ear offset (see Fig. 6.4). The larger the sphere, the closer the maxima are to the interaural axis. Additionally, the magnitude of these maxima increases for larger spheres. The positions and magnitudes of the minima for different subjects are almost constant.

---

[1]This chapter shows that it is more desirable to individualize the magnitude of an HRTF data set instead of the ILD. For this reason only 23 of 48 subjects are considered in the following which were already used for the published study of Bomhardt and Fels (2016).

[2]Shifting the positions of the ears on the sphere towards the back has the disadvantage that the ears are in the shadowing zone for frontal sources. For human heads, the ears are never in the shadowing zone for frontal sources.

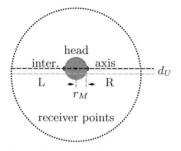
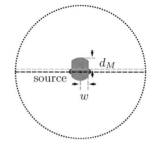

(a) A spherical head with an averaged radius r_M in the center.

(b) A head constructed from its width w, depth d_M and ear offset.

Figure 6.3.: Approximated heads: The two half circles (position of the ears) define the interaural axis which is displayed as a dark grey line. The receiver points are shifted by the ear offset d_U towards the back.

The falling edge from 98° to 180° is longer than the frontal one from 0° to 59°. This effect can be clearly assigned to the ear offset.

Determination of Numeric ILDs To investigate the influence of the pinna and head shape in comparison to a sphere, a numeric model with an approximated head and the individual pinna was constructed.

For that, a sphere was generated whose radius accorded with the depth of the head. Subsequently, this sphere was sliced by two planes, which were orthogonal to the interaural axis, at a distance of the head width (see Fig. 6.3). In comparison with an ellipsoid this geometry has the advantage that the sources (monopoles) at the ear canal entrance are not located in the shadowing zone for frontal sources[3]. According to the analytic setup, the receiver mesh was located around the origin in the horizontal plane. For the boundary element method, which was used for the simulation, this geometry was meshed by linear triangular elements with a maximum length of 5 mm. This element size is sufficient for the simulation up to 10 kHz without any spatial aliasing (Thompson and Pinsky, 1994). A rigid outer boundary condition as well as matched impedance (totally absorbent with an absorption coefficient of one) for the inner boundary condition was considered to avoid resonance effects inside the geometry.

To study the influence of the pinna, the three-dimensional ear models of these subjects were used. The models were adjusted on the sliced planes with the ear

---

[3]This applies only under the assumption that the ear offset is towards the back of the head.

offset of the respective subject. For sliced spheres with ears, the areas around the ears were meshed by smaller elements of a maximum 2.5 mm. This had the advantage that the delicate ear geometry was preserved. The receiver mesh and the boundary conditions remained the same.

Finally, the numeric ILDs can be determined by (6.1) from the simulated HRTFs and show two maxima between $0°$ and $180°$ at 2.5 kHz. Thus, the rear maximum is more pronounced than the frontal one (cf. Fig. 6.4). A further analysis of the dependency of the extrema and the head dimensions shows that the position of the rear maximum is linearly related to the depth of the head and the ear offset. The magnitudes of the maxima are mainly influenced by the depth of the head. The width of the head shows only a low correlation $|\rho| < 0.4$ with the extrema (Bomhardt and Fels, 2016).

Human ILDs   The human ILDs were calculated from the measured HRTF data sets by (6.1). Since the curve progression of the human ILD was previously exemplarily discussed at 2.5 kHz, the inter-subject differences in Fig. 6.4 should be investigated in the following.

All of the measured ILDs show two maxima and one minimum at 2.5 kHz (cf. Fig. 6.4). Although the frontal peak is often smaller than the rear one, there are several individual ILDs that show a larger rear peak. However, the inter-subject fluctuations of the rear peak are larger than the ones of the frontal peak. Assessing the correlation between the extrema and the head dimensions shows low correlation coefficients $|\rho| < 0.5$. One reason for this could be the low spatial resolution of $5°$ which is not sufficient for the determination of the maximum. Furthermore, the notches of the HRTF, which cause the maxima of the ILD, can be affected by a low SNR. Another reason is the three-dimensionally shaped head which is reduced to several one-dimensional measurements.

Anyhow, a weak correlation $0.1 < |\rho| < 0.3$ can be found between the depth of the head and the location of the rear maximum (frequency-averaged correlation coefficient between 1 kHz and 5 kHz of $\rho \approx 0.24$). The frequency-averaged correlation coefficients of the head width and the extrema (location and magnitude of the frontal and rear maximum) are below $\rho < 0.2$ (Bomhardt and Fels, 2016). It is assumed that the magnitudes of the maxima are not distinctly related to the head sizes due to sampling, SNR and head shape. Consequently, the correlation coefficients of the human ILDs are lower than for the analytic and numeric ones.

Comparison of ILD Cues   From the comparison of analytic, numeric and human ILDs in Fig. 6.4 it can be observed that the number of maxima and minima corresponds in all three cases at 2.5 kHz.

The maxima of the spherical ILDs are in general smaller and further apart than

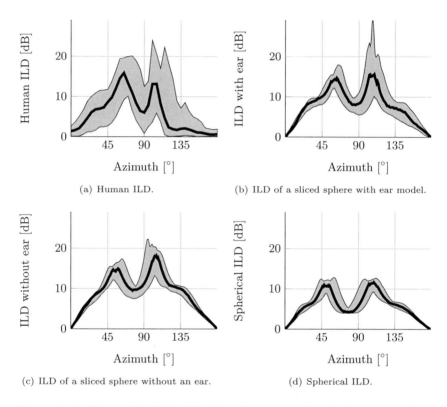

(a) Human ILD.

(b) ILD of a sliced sphere with ear model.

(c) ILD of a sliced sphere without an ear.

(d) Spherical ILD.

Figure 6.4.: The median ILDs (black lines) are plotted in the horizontal plane against the azimuth angle as absolute values. The grey areas mark the ventiles of the 23 subjects at 2.5 kHz.

the measured ones. Especially the frontal peak is $20°$ further to the front than the human one. Considering the depth of the head, these maxima are shifted towards the interaural axis. A relation between the position of this maximum and the depth of the head as well as the ear offset can be observed for all ILDs shown in Fig. 6.4.

The minimum can be found close to $90°$ for all considered subjects. The magnitudes at the minima of the numeric and human ILDs are in a comparable range while those of the sphere are several decibels lower.

The positions of the rear maxima are well aligned. However, the spherical rear peak is much smaller than the human one. The shadowing effect of the sliced sphere amplifies the rear peak. Comparing the curve progressions of the ILDs with and without ears, the pinna influences the steepness of the rear peak.

The shape of the head, which differs from the sliced sphere, influences the curve progression between $110°$ and $180°$. The magnitude maxima of the measured ILDs are smaller than the modeled ones.

To recap, the difference between the numeric ILD with pinna and the human ILD can be traced back to the deviating head shape and the orientation of the pinna. Both factors are not included in the current investigation.

6.3. Modeling of the Interaural Level Difference

Since spherical approximations of the human head are very common for the ITD estimation, the analytic spherical solution (Kuhn, 1977) is also appropriate as a low frequency approximation for the ILD. The comparison between the spherical and human ILD suggested that the human ILD is very similar to the spherical ILD below 1 kHz (Brungart and Rabinowitz, 1999). However, the asymmetry of the head, the ear position and the shoulder reflection cannot be considered by this approach. Additionally, it has to be remarked that humans do not make use of the ILD below 2 kHz (Rayleigh, 1907; Kulkarni et al., 1999) which is the reason why an adaption by a sphere as for the ITD is not appropriate: If the maxima of the spherical ILDs are considered, it turns out that they are less pronounced than the human ones. Furthermore, if the frequency exceeds 5 kHz, the ILDs differ in terms of number of maxima and amplitudes. Consequently, a spherical approximation is absolutely unsuitable. The differences between both result in changes to the perceived direction of the sound source (Phillips and Hall, 2005). These differences are caused by irregularities of the contralateral ear side which enable the localization on the *Cones of Confusion*. Losing such asymmetries, localization errors will rise (Carlile and Pralong, 1994).

Besides the analytic approximation of the ILD by a sphere for low frequencies, empirical methods using the superposition of sine functions by anthropometric dimensions provides an estimated ILD which reduces front-back confusions significantly (Watanabe et al., 2007).

Instead of modeling the ILD, it is also appropriate to model the HRTF since the characteristic cues of the ILDs are caused by the shadowing effect and the directivity of the averted pinna. Therefore, the individualization of the ILD can be applied indirectly by the individualization of the HRTF. Several methods and models as introduced in Chapter 3 are already known and are investigated in the following chapter in detail.

7

Spectral Cues of Head-Related Transfer Functions

The high frequent complex-valued spectrum of the HRTFs is mainly influenced by the shadowing effect of the head and the directional characteristic of the pinna. To derive spectral features as resonances or destructive interferences, the spectrum is described by its magnitude and phase. A resonance can be monitored by a direction-independent maximum of the magnitude while a destructive interference can be observed by a direction-dependent narrow-band notch. Both, resonance maximum and notch (cf. Iida et al. (2007) or Takemoto et al. (2012)), are important for the localization at frequencies above $2\,\mathrm{kHz}$ (Kulkarni et al., 1999). Since the ITD plays a major role for lower frequencies, the focus of the current chapter is on these spectral cues.

Plotting the HRTFs of different subjects and directions in Fig. 7.1 shows that these features are subject-dependent and directional (Møller et al., 1995b). At frequencies below the first resonance, the magnitudes vary less than at the ones above $5\,\mathrm{kHz}$ (Shaw and Teranishi, 1968). The resonance maximum around $5\,\mathrm{kHz}$ becomes smaller for contralateral HRTFs than for direction $(\theta, \varphi) = (0°, 270°)$ due to the shadowing effect. At higher frequencies, deep notches in the magnitudes can be observed which vary with frequency and direction (Raykar et al., 2005; Iida et al., 2007; Takemoto et al., 2012; Spagnol et al., 2013). Apart from these notches, narrow-band maxima can also be found at higher frequencies for frontal and ipsilateral HRTFs. The interference effects of the pinna are investigated subject-dependently in Section 7.1. Therefore, approaches are presented which enable the detection of such effects. Later in Section 7.6, these approaches are used to compare measured and estimated HRTFs. Furthermore, the symmetry of the ears and HRTFs are of interest since this symmetry reduces the measurement effort of anthropometric dimensions (cf. Section 7.3).

Since the direction-dependent notches depend on the anthropometric dimensions of the subject and the resonance frequency provides subject-dependent differences of the ILD, two different approaches to adapt an HRTF data set individually are introduced in this chapter.

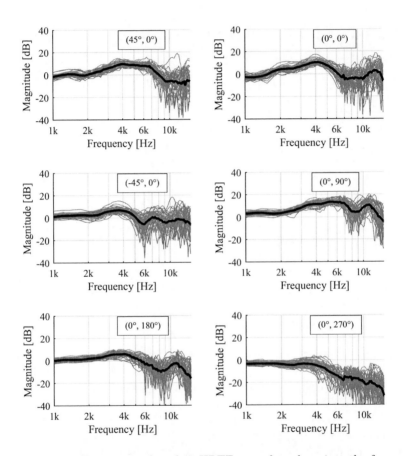

Figure 7.1.: The magnitudes of 47 HRTFs are plotted against the frequency for different directions (θ, φ). The black line marks the mean magnitude. While the left column of plots shows HRTFs in the median plane, the right column shows HRTFs in the horizontal plane for the left ear.

The first one uses a factor to scale the frequency vector which was introduced by Middlebrooks (1999a). This approach provides the opportunity to shift spectral cues in the frequency-domain. The optimal scaling factor for the HRTF to be adapted is determined by minimizing the error between the scaled HRTF and the individual HRTF (see next Section 7.2 for details). However, the existing individual HRTF data set already provides the best possibility for the binaural reproduction for which an adaption of a foreign data set is pointless. If the individual HRTF data set does not exist, the scaling factor for HRTF data sets has to be determined by other features than anthropometric dimensions. Based on the link between the anthropometric dimensions and the frequency-dependent spectral cues (cf. Section 7.1 and Kistler and Wightman (1992), Jin et al. (2000), and Ramos and Tommansini (2014)), the scaling factor can be expressed by a linear combination of dimensions. The localization performance with such a scaled HRTF is not as good as it would be with individual HRTFs but, however, better than with non-individual HRTFs (Middlebrooks, 1999b). The approach of Middlebrooks (1999a) was taken into account in this thesis to use in combination with the present database (Bomhardt et al., 2016a) and compare it to the empirical approach using PCA. A detailed description with further information about this adaption and its performance is given in Section 7.4.

The second approach uses principal components with direction- and subject-dependent weighting scores to reconstruct or estimate an HRTF data set (cf. Section 2.5.3 and Kistler and Wightman (1992), Jin et al. (2000), Inoue et al. (2005), Nishino et al. (2007), Hugeng and Gunawan (2010), and Ramos and Tommansini (2014)). After a general introduction for applying the PCA to HRTFs in Section 7.5, the complex-valued and real-valued reconstruction by principal components with their weighting scores is analyzed. Subsequently, the weighting scores are expressed by anthropometric features to create an individualized HRTF data set.

Finally, the mismatch of the spectral cues between the original and adapted HRTFs and the total adaption error is compared for both approaches.

7.1. Interference Effects of the Pinna

As already discussed in Section 2.1.2, humans use interferences as resonances or the destructive superposition of waves to localize sources on the *Cones of Confusion*. Therefore, these physical effects should be studied for the assessment of the quality of individualized HRTFs.

In this section, the interference effects of the pinna are briefly explained by modes of the simulated sound pressure on the ear surface. Subsequently, approaches to

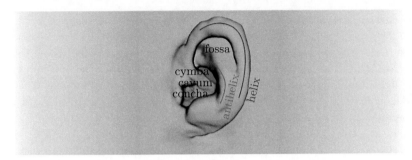

Figure 7.2.: The geometry of the ear can be described by the marked following parts: The cavity of the cavum concha, the cavity of the cymba, the cavity of the fossa, and the rims of the helix and antihelix.

detect the resonances and destructive interferences above $2\,\text{kHz}$[1] from HRTFs are presented.

The resonances can be adequately observed by measurement of the sound pressure at different points of the external ear (Shaw and Teranishi, 1968; Mokhtari et al., 2015) or a boundary element method simulation of the outer ear with a source inside the ear canal equally due to the reciprocity (Fahy, 1995; Katz, 2001).

In this thesis, the boundary element method is used to show the resonance effects at the pinna. For the simulation, the ear canal was closed by a surface approximately 5 mm behind the deepest point of the cavum concha. The source was located at the center on this surface. The surface around the outer ear was limited by a circle (see Fig. 7.3) and a perfectly matched layer (artificial absorbing layer) to avoid sound propagation behind the ear. A rigid outer boundary condition was used for the outer ear. The ear model was meshed with linear triangular elements (maximum edge length 2.5 mm). Consequently, no spatial aliasing was expected up to 10 kHz (Thompson and Pinsky, 1994). The sound pressure at the surfaces of the left ear in Fig. 7.3 was subsequently determined by the boundary element method. Since the absolute sound pressure level depends on the volume velocity of the monopole inside the ear canal, the relative sound pressure change at the outer ear is discussed in the following.

A resonance can be observed by a standing wave. This standing wave has often clearly defined sound pressure maxima and minima dependent on the geome-

---

[1] The shoulder reflection can be observed as a magnitude minimum in the frequency-domain or as a delayed reflection in the time-domain. Since the distance from the ear to the shoulder is large compared to the ear dimensions, the shoulder reflection is detectable as a less pronounced minimum between 1 and 2 kHz. Based on this fact, it is not considered for the assessment of the quality of individualized HRTFs in this chapter.

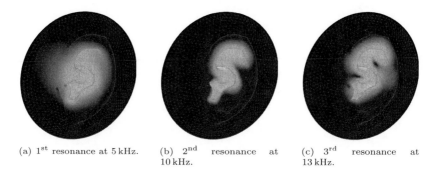

(a) 1st resonance at $5\,\text{kHz}$.   (b) 2nd resonance at $10\,\text{kHz}$.   (c) 3rd resonance at $13\,\text{kHz}$.

Figure 7.3.: Simulation results for the left ear of subject #17 (randomly chosen) are depicted for the first three resonance frequencies. The color marks the sound pressure level (the brighter the color, the higher the level). The plots cover a range of $30\,\text{dB}$.

try and the boundary condition for a specific frequency. Consequently, such a resonance can be monitored by a local maximum in the spectrum. The sound pressure level rises at specific areas and frequencies as in Fig. 7.4 and drops again afterwards. The higher the frequency, when such a sound pressure level maximum occurs, the smaller are the geometry dimensions of the outer ear which are affected. For most of the subjects three of these maxima can be observed in the range up to $15\,\text{kHz}$. The first three resonances of the left ear of randomly chosen subject #17 are shown in Fig. 7.3. The first mode shows a significantly higher sound pressure level in the complete cavum concha. The sound pressure level of the second one is also higher in the cavum concha than elsewhere. In comparison with the first mode, the level of the second one is lower at the rim between the cavum concha and cymba and higher at the fossa. The third mode of the ear shows a similar spatial distribution of the sound pressure level as the second one. Only inside the cavum concha, two minima can be observed. Therefore, it can be summarized for this subject that these three modes are mainly influenced by the fossa, cavum concha and cymba, which is in line with the study of Shaw and Teranishi (1968).

To investigate the resonances frequency-dependently, four different points are chosen on the basis of the shape of the standing waves in Fig. 7.3: One is located in the middle of the cavum concha, two at the rim between the cavum concha and cymba and one inside the cymba (see Fig. 7.4). From the sound pressure level at these points, the already depicted resonances can be detected by a maximum of the sound pressure level. Consequently, the first resonance around $5\,\text{kHz}$ for

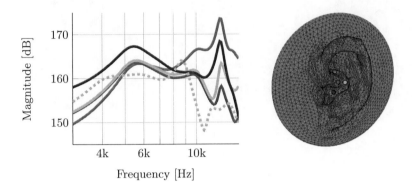

Figure 7.4.: The resonances are inside the cavum concha of subject #17 and are studied at the four different positions by their magnitude frequency-dependence. The dotted line marks the sound pressure level of subject #43 at the rim inside the cavum concha.

subject #17 can be adequately observed at all chosen points in Fig. 7.4 and is consistent with the findings of Shaw and Teranishi (1968). Nevertheless, the ears of some subjects, for example subject #43, do not show the first maximum at the point inside the cavity of the concha. For those, the sound pressure level increases steadily up to 8 kHz. After that, a clearly defined maximum can be found between 8 and 10 kHz (cf. dotted line in Fig. 7.4). Here, it is assumed that the first and second resonance are superimposed where the first resonance maximum cannot be detected.

If the first maximum is not as clear as for subject #17, the second resonance can be clearly observed between 8 and 10 kHz. The third resonance around 11 kHz is often observable by a rise of 10 dB sound pressure level. The rim inside the cavum plays an important role for this resonance because the sound pressure level drops significantly at the rim (see Fig. 7.3 and 7.4).

The destructive interferences cannot be detected from the sound pressure level inside the cavum concha because the direct incident wave is superposed direction-dependently by the reflection on the helix or antihelix. Consequently, they have to be detected from the HRTFs using either a tracking algorithm (Spagnol et al., 2013) or minimum detection (Raykar et al., 2005). In Section 7.1.2 a tracking algorithm for these notches is presented (Bomhardt, 2016; Bomhardt and Fels, 2017) which determines the notches for ipsilateral HRTFs.

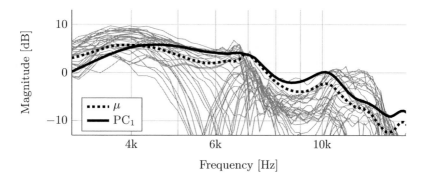

Figure 7.5.: The mean μ of all transfer functions and the scaled first principal component PC_1 of a randomly chosen data set #30 are plotted. Additionally, the magnitudes of the HRTFs in the median plane are shown as thin lines.

7.1.1. Detection of Resonances from Head-Related Transfer Functions

If the resonances of the outer ear are not superimposed by other resonances or destructive interferences, they are observable by frequency- and direction-independent maxima. Therefore, they should be detectable by the mean over all transfer functions (Mokhtari et al., 2015; Bomhardt and Fels, 2017) or the first principle component which contains the largest variances.

Peak Detection Using the Averaged HRTF   To calculate the mean μ over all HRTFs of one data set, magnitudes of the HRTFs are averaged (cf. Fig. 7.5 with an averaged HRTF μ). The maximum of this mean μ between 3 and 6 kHz is sometimes influenced by a notch or measurement noise (Takemoto et al., 2012), for which reason spatial averaging provides a more robust detection of the first peak (Mokhtari et al., 2015; Bomhardt and Fels, 2017). Applying the proposed peak detection on the 48 HRTF data sets of the present database (Bomhardt et al., 2016a), the first resonance maxima of the means μ can be found at 4.3 $\pm$ 0.7 kHz (mean and standard deviation). The magnitudes of these maxima amount to 7 $\pm$ 2 dB. The study of Mokhtari et al. (2015), which investigated HRTF data sets of the CIPIC database (Algazi et al., 2001c), determined resonance frequencies between 3.9 and 4.9 kHz with a mean value of 4.5 $\pm$ 0.3 kHz. While the mean value is comparable to the one of the current database, the standard deviation is larger. One reason could be the larger number of subjects of the present database (25 vs. 48 subjects). The magnitude of the maxima of Mokhtari et al. (2015) is slightly higher than in the present study and amounts to a mean of 11 ± 2 dB. However, the mean μ of one HRTF data set can be influenced by the number and position of measured HRTFs.

Since the first pinna notch often appears direction-dependent in the range of the second resonance, this resonance is often superimposed by this notch. Consequently, it is very challenging to detect from the mean μ (cf. Fig. 7.5).

Above 10 kHz often a well-defined maximum of the mean μ can be found. This maximum corresponds to the third resonance since the sound pressure rises significantly inside the cavum concha and cymba as shown for randomly chosen subjects in Fig. 7.4. This maxima can be found at frequencies of 11.1 ± 1.3 kHz with magnitudes of 1 ± 3 dB for the database at hand.

Neither the second nor third resonance are studied by Mokhtari et al. (2015).

Peak Detection Using the First Principal Component   The transfer functions of all data sets of the present database show a maximum which is clearly pronounced for ipsilateral HRTFs below 10 kHz (cf. Fig. 7.1). The maximum of the third resonance above 10 kHz of these HRTFs has a greater fluctuation due to the higher frequency and more complex directivity of the ear. Therefore, it is reasonable to detect this third resonance by the first principal component PC_1 which covers such variances. Remarkably, the lower first resonance frequency can be observed from the first principal component PC_1 by a wide maximum (cf. Fig. 7.5).

The first principal component PC_1 is calculated by (2.16) from the centered absolute value of the HRTF data set under investigation. Subsequently, only variances of the first resonance can be observed by the principal component PC_1 due to the zero-mean input data. These variances are mainly introduced by the lower magnitudes of the contralateral HRTFs and the first pinna notch which interrupts the first resonance above 6 kHz. For this reason, the first maximum above 3 kHz is wider than the maximum of the mean μ (cf. Fig. 7.5).

The detected first local magnitude maxima of the first principal component PC_1 are located at frequencies of 5.5 ± 1.0 kHz for the present database. These maxima are significantly higher than the detected maxima of the mean μ due to the minor influence of the first pinna notch.

Since the principal components are eigenvectors, they are scalable. Consequently, the magnitude of the maximum is not meaningful.

The second maximum is not significant enough to be robustly detected. However, the third resonance above 10 kHz is observable (cf. Fig. 7.5). The detected third resonance frequencies of the database are found at 11.2 ± 1.6 kHz which are similar to the ones of the mean μ.

7.1.2. Detection of Destructive Interferences from Head-Related Transfer Functions

Although destructive interferences, which are observed as notches in the HRTFs, are relevant on all *Cones of Confusion*, they are mostly studied in the median plane (Raykar et al., 2005; Spagnol et al., 2013). In the following, a tracking and minimum detection strategy is presented which detects these pinna notches for most of the ipsilateral HRTFs using a local minimum search in combination with a Kalman filter (Bomhardt, 2016; Bomhardt and Fels, 2017). The shoulder reflection is only observable by a slightly pronounced minimum between 1 and 2 kHz in the frequency-domain, for which reason it is not considered in this approach.

To observe the notches of an HRTF data set, a local minimum detection is applied on every transfer function. For the first pinna notch, all minima smaller than 0 dB are detected between 4 kHz and 11 kHz. To find related notches in the transfer functions, which are additionally disturbed by measurement noise, a Kalman filter is applied. In the present case, this Kalman filter is exploited as a tracking algorithm in a spherical slice over all elevation angles for a specific azimuth angle. Such a slice is shown in Fig. 7.6 where the notch can be observed for an elevation angle from $-60°$ up to $30°$. Above, the notch is located at frequencies larger than 10 kHz and is often affected by resonances and measurement noise. Therefore, the initial positions $x_0 = [\theta_0 \quad f_0]^T$ for the filter are chosen around $\theta_0 = -60°$ and $f_0 = 6$ kHz where the notch can be easily observed.

Considering that the position of the notch f_k will be shifted to higher frequencies for an increasing elevation angle θ_k, the next notch position x_{k+1} can be estimated in relation to the previous one x_k with the help of an underlying state transition model. The uncertainties, which occur due to measurement noise or resonances, are already considered as process noise by the Kalman filter. Further details of the tracking procedure can be found in the Appendix A.

In summary, the estimation procedure by the Kalman filter for the notch detection consists of three steps:

1. The prediction of the next measurement point.

2. A nearest neighbor search for the closest point in relation to the predicted point.

3. The correction of the model by the closest point.

As observed in Fig. 7.6, local minima are also detected in the upper hemisphere which seems to be frequency-independent and therefore rejected for tracking. The proposed tracking strategy was applied on the 48 HRTF data sets of the present database (Bomhardt et al., 2016a). For most of the subjects the eleva-

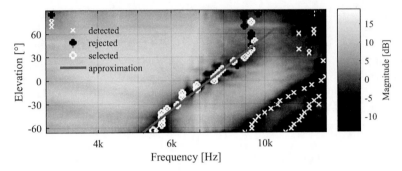

Figure 7.6.: The ipsilateral HRTFs of data set #30 are plotted frequency- and elevation-angle-dependently for azimuth angle $\varphi = 60°$. The detected minima are marked by crosses, the related notches (Kalman filter: first pinna notch) are marked by open circles, and the rejected minima are marked by black filled circles. The line is estimated by the detected minima.

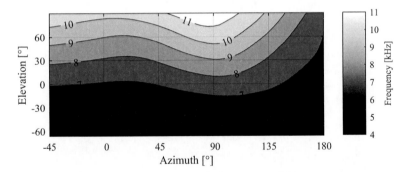

Figure 7.7.: The estimated first pinna notches of data set #30 are plotted over the azimuth and elevation angle (ipsilateral HRTFs only). The scale marks the frequency of the notch.

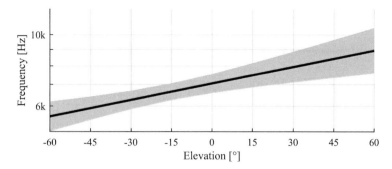

Figure 7.8.: The dark line is the mean notch frequency for the corresponding elevation angle and the gray area marks the standard deviation over all data sets of the present database in the median plane (left ear).

tion angle of the first and often also the second notch increase linearly with a logarithmic frequency. Therefore, the notches are fitted by first order polynomial $\theta_n = m \cdot \log(f) + n$.

The detected first pinna notches of a randomly chosen data set #30 are shown in Fig. 7.8. From this representation, it is obvious that the notch frequency is not only elevation-dependent but also azimuth-dependent. For frontal and rear directions, it increases from 6 to 11 kHz in the range of $-60° \leq \theta \leq 60°$ while for lateral directions it increases stronger. Nevertheless, when contralateral HRTFs were investigated, the shadowing effect and the measurement noise disturbed the notch detection, therefore the estimation was not reliable anymore in this region. The Fig. 7.8 shows in addition that the notch is subject-dependent. The larger standard deviation at higher frequencies and at higher elevation angles respectively is caused by the short wave lengths and the influence of resonances of the cavum concha.

Raykar et al. (2005) detected the notches in the time-domain but did not assign them for which reason it cannot be compared with the present approach. The study of Takemoto et al. (2012) investigated the notches in the median plane by numeric simulations. Since this study is focused on the mechanism which generates the notches, only four subjects were considered. The notches of these subjects were between 5 and 8 kHz in a range of $-30° \leq \theta \leq 50°$ in the median plane. Spagnol et al. (2013) used 20 subjects from the CIPIC database to detect the first three notches by a tracking algorithm from PRTFs. The first notch increased monotonously in the range $-45° \leq \theta \leq 45°$ from 6 ± 1.5 kHz to 9 ± 1.0 kHz. The present database with the Kalman tracking algorithm achieved similar results in this range: $5.9 \pm 1.1\,\text{kHz}$ at $(\theta, \phi) = (-45°, 0°)$ and $8.4 \pm 1.1\,\text{kHz}$ at $(45°, 0°)$ (cf. Fig. 7.8).

7.2. Evaluation Standards of Spectral Differences

Section 7.1 already introduces approaches to identify spectral cues as resonances and destructive interferences of the HRTF. Consequently, the detected spectral cues can be used to investigate the difference between two HRTF data sets or the symmetry of HRTFs. The current chapter describes different procedures to compare complete HRTF data sets rather than particular cues (Middlebrooks, 1999a; Richter et al., 2016).

First of all, the inter-subject spectral difference (ISSD_{dir}) of Middlebrooks (1999a) is introduced. The difference between two HRTF data sets is characterized by the variance of the frequency-dependent ratio of both. In this case, the variance allows a closer look into the subject-dependent difference than the mean since this can also be influenced by level offsets. Subsequently, the resulting variance of this ratio has to be determined for each direction i of the data sets to derive the mean variance

$$\text{ISSD}_{dir} = \frac{1}{n_{dir}} \sum_{i=1}^{n_{dir}} \text{Var} \left(20 \log_{10} \frac{|\text{HRTF}_{1,i}(f_j)|}{|\text{HRTF}_{2,i}(f_j)|} \right). \qquad (7.1)$$

In contrast to the ISSD_{dir} by Middlebrooks (1999a), which was calculated for 64 frequency bands in the range from 3.7 to 12.9\,kHz, the current ISSD_{dir} is calculated for each frequency bin f_j between 1 and 13\,kHz. The ISSD_{dir}s at hand are larger than the ones by Middlebrooks since the current data sets have a different spatial sampling grid and the higher number of directions. Dependency on the sampling can only be reduced if both studies take surface weights into account (cf. Section 5.4). However, surface-weighted ISSD_{dir}s are not significantly different to the one of (7.1). Consequently, the weights are neglected in the following. Additionally, the study by Middlebrooks used DTFs instead of HRTFs for the optimization[2].

A different frequency-dependent measure by Richter et al. (2016)

$$\text{ISSD}(f_j) = \sigma \left(20 \log_{10} \frac{|\text{HRTF}_{1,i}(f_j)|}{|\text{HRTF}_{2,i}(f_j)|} \right) \qquad (7.2)$$

---

[2]Nevertheless, the scaling factors for the DTFs and HRTFs are very similar for which reason the HRTFs, which preserve the phase, are preferred (cf. Section 7.4).

calculates the standard deviation of the directions for each frequency bin f_j. The sum over all frequencies

$$\text{ISSD}_f = \frac{1}{n_f} \sum_{j=1}^{n_f} \text{ISSD}(f_j)^2 \qquad (7.3)$$

results in a representation similar to (7.1).

To summarize, the measure ISSD_{dir} (7.3) weighted fluctuations of the frequency more strongly, while the measure measure ISSD_f (7.1) is more robust to direction-dependent spectral differences. The inter-ear differences (IESDs) of the present database (Bomhardt et al., 2016a) are presented in Section 7.3 and inter-subject differences (ISSDs) are discussed in Sections 7.3.2 as well as 7.4.2.

7.3. Symmetry of the Ears

As shown in Fig. 7.9, both ears are often very similar but differ in detail. To individualize HRTFs by their anthropometric dimensions, it would be desirable to reduce the measurement procedure under the assumption of symmetric ears (cf. Bomhardt and Fels (2017)).

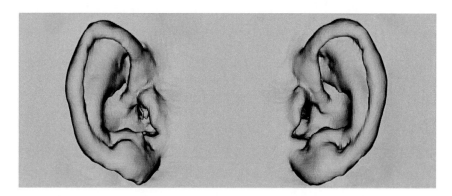

Figure 7.9.: The three-dimensional ear models (right and left) of subject #33 from the present database (Bomhardt et al., 2016a).

Based on this fact, this section deals with the difference between right and left ears of the present database (Bomhardt et al., 2016a). It is thematically split by the analysis of the measured anthropometric dimensions and the analysis of the observable interference effects.

	d_1	d_2	d_3	d_4	d_5	d_6	d_7
Δ_{abs}	1	1	1	2	2	2	1
Δ_{max}	4	4	6	6	5	5	4
Δ_{rel}	7	14	8	9	3	4	23
ρ_{LR}	0.6	0.6	0.8	0.8	0.9	0.8	0.3

Table 7.1.: Comparison of the dimensions of the right and left ear with the difference of both. Δ_{abs} and Δ_{max} are given in millimeters and Δ_{rel} is expressed in percentage.

7.3.1. Symmetry of Anthropometric Dimensions

The collected one-dimensional anthropometric dimensions from the present database according to the CIPIC specifications (Algazi et al., 2001c) can be found in Table 4.3 and Fig. 4.4. To briefly summarize this table: The largest dimensions were the ear height d_5 and width d_6 which varied between 30 and 74 mm. The fossa height d_4, the cavum concha height d_1 and width d_3 were between 13 and 28 mm large. The smallest dimensions were the cymba height d_2 and the intertragal incisure width d_7 with 4 to 11 mm.

However, repeated measurements of randomly chosen ears showed that the manually-detected measurement points caused measurement uncertainties of about 1 mm. It is assumed that the complex shape of the individually character-istic shape of the ear and the definition of the individual measurement points led to these deviations.

In the following, the anthropometric dimensions of the pinna are split into right (R) and left (L) ear dimensions to compare them. The difference between the right and left ear dimensions is evaluated by the difference of the anthropometric dimensions of both ears $|d_L - d_R|$ (cf. Table 7.1). Therefore, the averaged absolute, relative and maximum difference

$$\Delta_{abs} = |d_L - d_R| \tag{7.4}$$

$$\Delta_{max} = \max_d |d_L - d_R| \tag{7.5}$$

$$\Delta_{rel} = |d_L - d_R|/d_L \cdot 100 \tag{7.6}$$

were calculated.

The mean absolute difference Δ_{abs} between right and left dimensions varies between 1 and 2 mm. Large dimensions, such as d_4, d_5 and d_6, show larger deviations than smaller ones. The same applies to the maximum differences Δ_{max} and relative differences Δ_{rel}. In addition, larger deviations tend to result in a lower correlation coefficient ρ_{LR} between the right and left ear dimensions

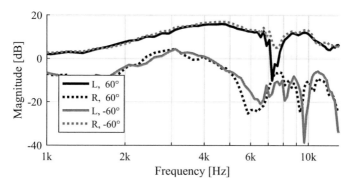

Figure 7.10.: Measured HRTFs of data set #33 at symmetric positons $(\theta, \varphi) = (0°, \pm60°)$ for the right (R) and left (L) ear.

(Zhong et al., 2013). However, the larger deviations cannot be directly assigned to measurement uncertainties or deviating ear geometry.

7.3.2. Symmetry of Head-Related Transfer Functions

Neither the ears (cf. Fig. 7.9), nor the HRTFs are completely symmetric (cf. Fig. 7.10). The difference between symmetric directions $\pm\varphi$ can be caused by the anthropometric dimensions or uncertainties such as measurement noise or misalignment of the subject (cf. Section 5.2.1).

In the following, inter-ear differences of the data sets are investigated as well as the first resonance and notch.

Inter-Ear Difference To investigate the asymmetry of a complete HRTF data set, the ISSD_{dir} (7.1) is modified to express the inter-ear spectral difference

$$\mathrm{IESD}_{dir} = \frac{1}{n_{dir}} \sum_{i=1}^{n_{dir}} \mathrm{Var}\left(20\log_{10}\frac{|\mathrm{HRTF}_{L,i}(f_j)|}{|\mathrm{HRTF}_{R,i}(f_j)|}\right). \quad (7.7)$$

For this purpose, the HRTF data set is split into HRTFs of the right ear HRTF_R and left ear HRTF_L. Additionally, the symmetric directions $\pm\varphi$, have to be mirrored by $\varphi_M = 360° - \varphi_R$.

Considering all subjects of the present database, the deviation between right and left ear amounts to $19 \pm 7\,\mathrm{dB}^2$ between 1 and 13 kHz. This is significantly lower than the subject-averaged ISSD_{dir} of $34 \pm 9\,\mathrm{dB}^2$.

Furthermore, the frequency-dependent measure ISSD (7.2) is also modified as

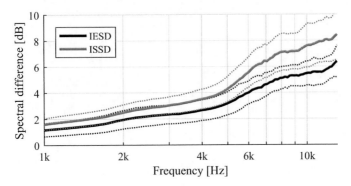

Figure 7.11.: The solid line of the frequency-dependent inter-ear and inter-subject difference marks the mean of all subjects. The dotted lines show the corresponding standard deviations.

inter-ear difference

$$\text{IESD}\,(f_j) = \sigma\left(20\log_{10}\frac{|\text{HRTF}_{L,i}(f_j)|}{|\text{HRTF}_{R,i}(f_j)|}\right) \tag{7.8}$$

to investigate the deviations between the right and left HRTFs frequency-dependently.

The subject-averaged IESD features an increasing mismatch towards higher frequencies (see Fig. 7.11) which accords with the deviations of individual HRTFs in Fig. 7.1. At frequencies below 5 kHz, the subject-averaged IESD is about 2 dB and increases especially in the range between 6 and 9 kHz. This increasing difference can be ascribed to the notches of the HRTFs.

Comparing the frequency-dependent IESD with the ISSD in Fig. 7.11, the spectral difference between the ears is smaller than the difference between the subjects.

However, repeated HRTF measurements[3] of one subject showed comparable differences to the IESD$_{dir}$ or IESD.

Symmetry of the First Resonance   The first resonance frequency can be detected by an averaged HRTF μ of a data set as described in Section 7.1.1. To recap, the subject-averaged resonance frequency of the present database is 4.3 ± 0.7 kHz.

In the following, the resonance frequencies of the data sets were determined for both ears separately. It is assumed that the averaged human ears do not show

---

[3]Subjects #1 and #17 were measured four times on different days with a reconstructed measurement setup.

	Mean ± std.	Δ_{LR} (mean ± std.)	ρ_{LR}
f_R	4.3 ± 0.7 kHz	0 ± 0.3 kHz	0.9
$f_{N,10°}$	6.4 ± 0.5 kHz	0 ± 0.4 kHz	0.6
$f_{N,30°}$	6.5 ± 0.5 kHz	0 ± 0.4 kHz	0.7
$f_{N,60°}$	6.8 ± 0.6 kHz	-0.1 ± 0.3 kHz	0.8
$f_{N,70°}$	7.0 ± 0.6 kHz	0 ± 0.3 kHz	0.8

Table 7.2.: The resonance frequencies f_R as well as the notch frequencies f_N at different elevation angles θ in the horizontal plane are shown. In addition, the difference Δ_{LR} between the right and left ear as well as their correlation ρ_{LR} are listed.

any general deviation for one ear side which is confirmed by a small subject-averaged difference of 27 Hz. However, the standard deviation and the correlation coefficient ρ_{LR} provide the difference between both ears. The averaged deviation between the ears amounts to 0.3 kHz and both resonance frequencies show a strong correlation $\rho_{LR} = 0.9$.

Comparing the magnitudes of the first notch of the right and left ear shows a difference of 1.2 dB and a correlation coefficient of 0.8.

Symmetry of Destructive Interferences The difference between the notches can only be applied for specific directions. Therefore four symmetric directions $\varphi = \{\pm 10°, \pm 30°, \pm 60°, \pm 70°\}$ were randomly chosen in the horizontal plane[4]. Since the detection of the notches for rear or contralateral directions is difficult, only the ipsilateral HRTFs in the frontal hemisphere were considered. The notches of the noted directions were detected from the corresponding spherical slices and the polynomial fit as proposed in Section 7.1.2. From Table 7.2, it can be observed that the subject-averaged notch frequency rises for an increasing azimuth angle. This effect has already been monitored in Fig. 7.7 for a single subject. Considering the differences between the ipsilateral right and left ear notches in this Table 7.2, the maximum averaged difference amounts to -55 Hz so that no general difference is observed. The inter-ear deviations between the notches are between 0.3 and 0.4 kHz.

Notch Frequency and Related Anthropometric Measurement Error At the end of this section, the relationship between the notch frequency f_N, the anthropometric dimension l and the speed of sound c_0

$$f_N = \frac{c_0}{4l} \qquad (7.9)$$

---

[4]These directions were also used for the listening experiment with symmetric HRTFs in Section 7.5.3.

of Spagnol et al. (2013) is used to briefly discuss the required measurement accuracy for the anthropometric dimensions.

From the notch frequencies of the present database, the anthropometric dimension l can be calculated. Since this notch frequency is direction-dependent, the dimension varies between 10 and 15 mm for the first notch. It is assumed that the rim of the antihelix causes the first notch, therefore it is reasonable that this dimension is a little bit smaller than the concha width d_2.

Using the observed differences between the right and left ear of $\Delta f_N = 0.3\,\text{kHz}$ from Table 4.3, the corresponding deviations amount to 1 mm. As previously mentioned in this section, repeated measurements of the anthropometric dimensions already showed a deviation of 1 mm. For this reason, it can be summarized that it is very challenging to consider the anthropometric difference between the ears accurately.

7.4. Individualization of the Head-Related Transfer Function by Frequency Scaling

The individualization of an HRTF data set by frequency scaling is motivated by the observation that larger anthropometric dimensions will result in spectral cues at lower frequencies (Middlebrooks, 1999a,b). The spectral cues can be shifted to lower or higher frequencies using a scaling factor s. This scaling factor is multiplied by the frequency vector of the HRTFs

$$\text{HRTF}_s\,(f) = \text{HRTF}\,(s \cdot f)\,. \tag{7.10}$$

If the scaling factor is less than one, the cues will be shifted to lower frequencies and vice versa. The squeezed frequency vector results in a frequency shift if this vector is displayed logarithmically. The logarithmic approach of the frequency vector is very common in acoustics since the auditory system perceives the frequency logarithmically (Pikler, 1966). Consequently, the squeezing is called scaling in the following.

Applying this to discrete data, the sampling points have to be interpolated. Furthermore, the measured HRTFs have a limited bandwidth which forces a loss of information. For scaling factors larger than one, for instance, low frequent information will be lost which affects the localization less than the loss of high frequent information. In particular, the spectrum at low frequencies is very flat while at higher frequencies the magnitude fluctuates more intensely due to the influence of the shadowing effect and the pinna.

Subjective investigations from a listening experiment showed that the front-back errors increase when listening with an HRTF of a larger subject (Middlebrooks,

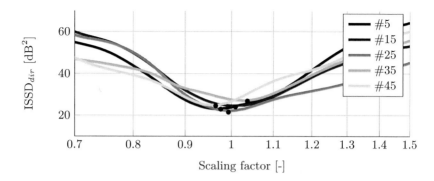

Figure 7.12.: The ISSD_{dir} is plotted against the scaling factor for five randomly chosen subjects and scaled HRTFs of subject #19. The minima are marked by black dots.

1999b). In the opposite case, when an HRTF of a smaller subject is used, the back-front errors increase. Subsequently, it is desirable to minimize these errors by finding the optimal scaling factor for the HRTFs to be individualized.

In the following sections, the adaption by an anthropometric scaling factor is described and accords in general with the proposed approach of Middlebrooks (1999a). In detail, the adaption of an existing HRTF data set by the anthropometric dimensions differs. At the end of this chapter scaled HRTFs are compared to estimated HRTFs from principal components by their spectral cues.

7.4.1. Optimal Scaling Factor

The optimal scaling factor between two different HRTF data sets can be found on the basis of the inter-subject spectral difference by a numeric minimum search (Middlebrooks, 1999a). Therefore, the range of the scaling factor has to be limited. Since the differences are introduced by anthropometric dimensions, the minimum will be detected for scaling factors between 0.5 and 2. The optimal scaling factors were determined in Fig. 7.12 for the five selected subjects and the HRTF data set #19. Plotting the ISSD_{dir} against the scaling factor shows a pronounced minimum close to $s = 1$ for each of the five subjects. Considering all 41 remaining HRTF data sets of the present database, the scaling factors of the adapted HRTF data set #19 range from $s = 0.84$ to 1.12.

Since the ISSD_{dir} for the optimal scaling factor measures the difference between the adapted and true HRTFs of a subject, the direct comparison of the adapted and individual HRTFs of subject #7 shows the differences more clearly (see Fig. 7.13). To adapt the HRTF data set of subject #19 for subject #7, the optimal

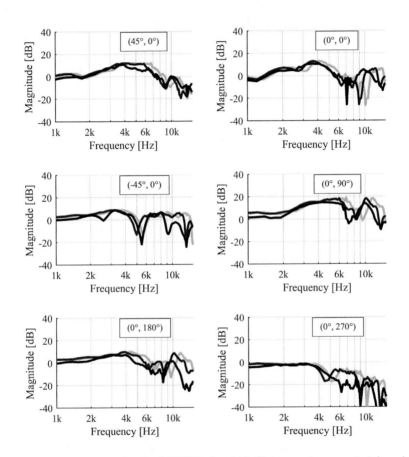

Figure 7.13.: The individual HRTFs for #19 (light gray) are scaled (gray) by the optimal scaling factor of $s = 0.88$ to match the HRTFs for #7 (black) at different directions (θ, φ).

scaling factor of $s = 0.88$ was used which shifted the spectral cues towards lower frequency so that they are better aligned with the individually measured HRTFs of subject #7.

The resonance frequency around 3.5 kHz of subject #7 is smaller than the one of subject #19. The scaling of the data set #19 minimizes this mismatch. While most of the notches do not match exactly, the second maximum fits very well in Fig. 7.13. However, deviations between the scaled HRTFs and the individual ones increase for contralateral HRTFs where the magnitude of the individual HRTF is lower than the one of scaled HRTF. Unfortunately, the magnitude cannot be adapted by frequency scaling therefore larger differences at frequencies above 10 kHz can be depicted.

Overall, the scaling reduces the ISSD_{dir} of the current HRTF data sets (#7 and #19) from 33 to 25 dB^2. This improvement of 24% is in line with the study by Middlebrooks (1999a). In general, the data set #19 achieves an averaged reduction from 28 ± 7 dB^2 to 24 ± 5 dB^2 for the data sets in the present database which is an averaged improvement of 4 dB^2. To determine this difference, every single data set was scaled by the optimal scaling factors of all other data sets.

Investigation on the inter-subject difference of all data sets in the present database showed an averaged improvement up to 5 dB^2 for an adapted data set compared to the original data sets. The analysis showed that some of the data sets differ significantly from most of the other ones so that scaling does not improve the ISSD_{dir} much. The best matching unscaled HRTF data sets (#7, #16, #19, #31 and #47) provide an averaged ISSD_{dir} of 29 dB^2. If they are used for scaling, the averaged ISSD_{dir} decreases to 24 dB^2. As a comparison, the ISSD_{dir} of the scaled data sets is only 5 dB^2 higher than the IESD of the database at hand (cf. Section 7.3).

As the frequency vector is scaled, the phase is scaled as well. This implies that the ITD is adapted by the scaling. However, this ITD is adapted by a criterion which considers frequencies above 1 kHz. Therefore, a separate adaption of the ITD after the frequency scaling by a minimum phase and a modeled ITD is reasonable. Further optimization by rotating the measurement grid around the interaural axis can reduce the error due to the orientation of the pinna (Middlebrooks, 1999a; Guillon et al., 2008).

7.4.2. Anthropometric Scaling Factor

In case where the HRTF data set of a subject is not available, the scaling factor can be expressed by its anthropometric dimensions. For this purpose, Middlebrooks (1999a) used the ratio between the head width and cavum concha height to determine the estimated scaling factor. This has the advantage that only the anthropometric dimensions of the subject as well as the one of the HRTF to be scaled has to be known.

In contrast to this, the current thesis studies the complete set of anthropometric dimensions of a subject to estimate the scaling factor. Assuming a linear relationship between the anthropometric dimensions α_i and the scaling factor s_j of subject #j, a regression analysis (Seber, G. A. F and Lee, 2003, pp. 1-12) can be used to calculate this scaling factor s_j by a linear combination of n_{anthro} anthropometric features $\alpha_{j,i}$ and regression coefficients $\beta_{j,i}$ (see Appendix B for further details)

$$\hat{s}_j \quad = \quad \beta_{j,0} + \sum_{i=1}^{n_{anthro}} \beta_{j,i}\, \alpha_{j,i}. \tag{7.11}$$

By applying the linear regression analysis to the present database[5] (Bomhardt et al., 2016a), the resulting estimated scaling factors show a good agreement with the optimal ones (correlation coefficient of 0.79). However, comparing the anthropometric scaled HRTFs of subject #19 with the other HRTFs of the present database, the averaged ISSD_{dir} amounts to $26 \pm 5\,\text{dB}^2$ and improves the accordance between the subjects by $3\,\text{dB}^2$. The comparison between the optimal scaled and original data is $5\,\text{dB}^2$.

Since it is very time-consuming to collect 13 anthropometric dimensions for the adaption process, the influence of each dimension on the estimated scaling factor is analyzed to reduce the number of required dimensions. Here, the regression coefficients β are considered which weight each dimension. The higher the coefficient, the higher the impact of the dimension on the scaling factor. So, the dimensions are ordered, ascending by their coefficients $|\beta|$: h, d_M, d_1, d_3, w, d_5, d_8, d_4, d_U, d_6, d_B, d_2 and d_7 (dimensions are specified in Fig 4.4). Reducing the input parameters to the six most influential β still shows a good agreement with the optimal scaling factors (correlation coefficient of 0.77) and differs only slightly from the one which includes all 13 anthropometric dimensions (correlation coefficient of 0.79).

Further analysis of the sign of the regression coefficients shows that heads with a large depth have a smaller estimated scaling factor. The same applies to the ear offset towards the back of the head and the intertragal incisure width d_7. The cymba concha height d_2 has the strongest positive impact, showing large subject-dependent fluctuations of the regression coefficient. Interestingly, the study by Fels and Vorländer (2009), which investigated the anthropometric

---

[5] Forty-seven subjects and 13 anthropometric dimensions were used since the HRTF data set of one subject has a lower spatial resolution than the others.

influence of the shoulder, head and pinna, showed similar results: Especially the depth of the head, the cavum concha width and height influences the HRTFs while the head height and the dimensions of the pinna play a minor role.

7.4.3. Frequency-Dependent Comparison of Scaling Factors

After the description of the optimal and estimated scaling factors in the last two sections, both should be compared by the frequency-dependent ISSD (7.2) in the current section. For this purpose, three different scaling factors are considered:

s: Based on the fact that the HRTF data set #19 is one of the best matching HRTF data sets (cf. Section 7.4.1), this data set was chosen for the adaption. The optimal scaling factors were determined by minimizing the ISSD_{dir} for each subject of the database.

$\hat{s}$: The estimated anthropometric scaling factor was derived from the optimal scaling factors $\mathbf{s}$ of subject #19 and the anthropometric dimensions of the subjects by (B.6).

s_f: As an alternative to the ISSD_{dir}, the optimal scaling factor s_f was determined by minimizing the ISSD_f which weights fluctuations of the frequency more strongly. The resulting $\mathbf{s}_f$ were calculated for HRTF data set #19.

The decreasing accordance between a scaled and individual HRTF data set at higher frequencies has already been shown in Fig. 7.13 which considers the true set #7, the scaled set #19 and the unscaled set #19. The same behavior can be found by analyzing the frequency-dependent ISSD in Fig. 7.14. Comparing the ISSD of the scaled HRTFs with the unscaled ones of subject #19, the scaling improves the accordance of the data sets between 4 and 12 kHz. The benefit of scaled HRTFs decreases at higher frequencies due to the fact that the scaling factor s is optimized in the range from 1 to 13 kHz. Using a different data set for the scaling will result in larger ISSDs since the set #19 is one of the best matching sets (cf. Section 7.4.1). Consequently, the improvement with a data set, which has a larger initial ISSD, will be higher.

Furthermore, the frequency-dependent ISSD in Fig. 7.14 shows that there is no significant difference between the optimal and estimated scaling factor (s and $\hat{s}$). Only a slight benefit of the optimal scaling factor between 9 and 11 kHz is visible which could have been caused by the notch-frequency.

The optimal scaling factor s_f, which was determined by minimizing ISSD_f,

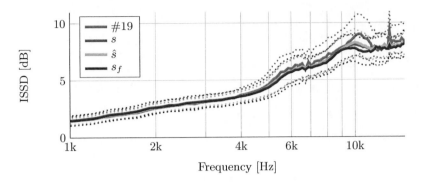

Figure 7.14.: The frequency-dependent ISSD of the HRTF data set #19 and the other sets of the database is plotted. The averaged difference is marked by a solid line and the standard deviation by dotted lines. Additionally, the differences between the scaled data sets of subject #19 and the true sets are shown.

achieves lower ISSDs in the frequency range between 6 and 12 kHz, therefore this criterion is an alternative to the ISSD_{dir}.

7.5. Individualization of Head-Related Transfer Functions by Principle Components

The principal component analysis (PCA) has already been introduced in Section 2.5.3 and can be utilized to decompose, reconstruct and individualize HRTF data sets. Its input data can be either HRIRs or HRTF data sets. An optimal reconstruction is feasible by considering the total number scores $\mathbf{W}$ and principle components $\mathbf{V}$ (see (2.17) and (2.16)). Neglecting the higher-order components and scores usually provides a sufficient approximation. A criterion to assess this approximation is the cumulative variance of the reconstructed data. A very common threshold is that 90% of the variance across the input data should be accounted for (Kistler and Wightman, 1992). The number of components which are necessary to fulfill this threshold varies between five and 12 PCs dependent on the input data (Kistler and Wightman, 1992; Jin et al., 2000; Ramos and Tommansini, 2014).

Most of the studies use magnitudes of the DTFs for the PCA which points to the fact that the phase is lost. Since human localization makes use of the phase especially at lower frequencies, the phase can be replicated by a minimum phase and an ITD (Kistler and Wightman, 1992; Jin et al., 2000) (cf. Section 2.5.4 and

5.2.3). However, the PCA can also be applied for the complex-valued spectrum what remains the ITD (Ramos and Tommansini, 2014). Due to the complex spectrum, a slightly higher number of PCs is necessary to reconstruct the data sets. Alternatively, the HRIRs can also be used as input data for the PCA whereby a similar number of components as for HRTFs is necessary to account for 90% of the variances (Hwang and Park, 2008; Hwang et al., 2008; Shin and Park, 2008; Fink and Ray, 2015).

In contrast to the mentioned studies, the present investigations are focused on the comparison of different individualization approaches of HRTFs. Subsequently, not only the averaged spectral differences are investigated (cf. Section 7.2) but also the quality of spectral cues of the individualized HRTFs (cf. Section 7.1). Furthermore, the reduction of necessary principal components by grouping, and the number of necessary anthropometric dimensions is discussed (see Sections 7.5.1 and 7.5.2). Finally, a listening experiment is presented which compares the different individualization types.

7.5.1. Reconstruction of the Spectrum

The current input data for the PCA has $n = 216\,576$ observations resulting from 2304 directions, two ears and 47 subjects as well as $p = 129$ frequency bins (Bomhardt et al., 2016a). The maximum number of eigenvectors $\mathbf{V}$ is then calculated from the rank of the input data matrix $\mathbf{H}$ which is in this case p. The corresponding weighting score matrix has a dimension of $n \times p$.

In the following paragraph examines the PCs and scores of a PCA on the basis of linear magnitude spectra (applying the PCA on the logarithmic magnitude spectra shows similar results to the linear one (Ramos and Tommansini, 2014)). Afterwards the approximated reconstruction of the HRTFs with less than p PCs is investigated for the real-valued and complex-valued spectra.

Magnitude Spectrum The PCA was applied on the linear magnitude of the HRTFs of the present database. In the following, the first components and their weights are investigated to show their spectral influence.

Since the first component PC_1 accounts for the largest variances of the input data, it shows a clear maximum around 5 kHz (see Fig. 7.15). This maximum is located slightly higher than the one of the direction-averaged HRTF μ (cf. Section 7.1.1). All higher-order PCs show at least one minimum and one maximum. The second and fifth PCs have their minimum around 8 kHz and cover the notches which appear at these frequencies (cf. Fig 7.1). In contrast to the first component PC_1, the third and fourth have their minimum around 4 kHz. Probably, these minima are useful to shift and amplify the first resonance subject-dependently. This

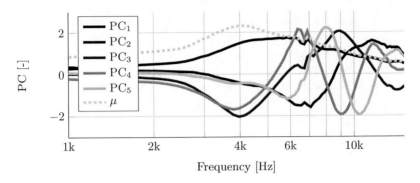

Figure 7.15.: The first five PCs are plotted against the frequency. Additionally, the averaged magnitude μ for all directions and subjects is shown (dashed line).

assumption can be supported by the analysis of the scores of both components which have low amplitudes and scatter subject-dependently. In general, the half width of the maxima and minima of the PCs decrease for high-order PCs so that these PCs can be interpreted as notch and band-pass filter. Although Shin and Park (2008) studied principal components in the time-domain, they also concluded that the inter-subject variations are influenced by the first three components.

The influence of the PCs at specific directions can be derived by a closer look at the scores in the horizontal and median plane in Fig. 7.16 and 7.17. The scores amplify the first component PC_1 at the ipsilateral ear and reduce its impact at the contralateral ear. The same behavior can be found in the curve progression of the HRTFs in the horizontal plane around the resonance frequency at 5 kHz in Fig. 7.1. Higher-order components have in general a low impact at the averted ear side. Their scores mainly influence the reconstructed HRTFs in the frontal hemisphere. If scores of the order of six and higher are considered, the subject-averaged scores are close to zero. Consequently, it can be assumed that they are responsible for the inter-subject difference.

In the median plane, the scores of the first component PC_1 show the largest amplitudes. They increase almost up to $30°$ and decrease again for elevations larger than $60°$. The inter-subject deviations of these scores are larger for upper than for lower elevations. Especially the scores of the second and fifth components, which have their maximum around 7 kHz, influence the reconstructed HRTFs between $-45°$ and $15°$. The scores of higher-order components show a larger impact on lower directions than on upper ones. However, their averaged scores fluctuate

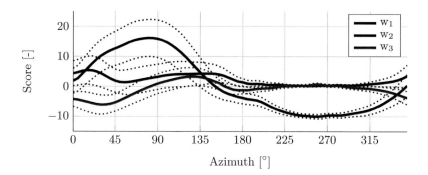

Figure 7.16.: For reasons of clarity, only the subject-averaged scores of the first three PCs are plotted against azimuth in the horizontal plane. The dotted lines mark the range of the standard deviation.

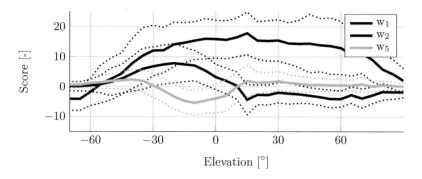

Figure 7.17.: For reasons of clarity, the most interesting subject-averaged scores of PC_1, PC_2 and PC_5 are plotted against elevation in the median plane. The dotted lines mark the range of the standard deviation.

around zero representing inter-subject differences. Although Hwang and Park (2008) calculated the PCs in the time-domain, they observed similar effects. They concluded that the first component PC_1 provides cues for the vertical perception whereby the inter-subject variations of the scores are larger in the frontal region than in the rear region. The scores of the second and third components provide cues for the front-back discrimination meanwhile the remaining ones provide subject-specific fluctuations[6].

Since the first five components and their weights have been investigated in detail, the number of necessary components for a sufficient reconstruction should be discussed now. In literature the required number of components is motivated by the threshold where the cumulative variance reaches 90% (Kistler and Wightman, 1992; Hwang and Park, 2008; Ramos and Tommansini, 2014). Dependent on the input data, the number of required components to fulfill this limit varies between five and six[7]. Considering the present database with 47 subjects, 2304 directions and 129 frequency bins as input data, the threshold is reached for six components.

In comparison with the original HRTFs, the reconstructed ones with six components show very similar curve progressions for the upper and ipsilateral directions in Fig. 7.18. This accordance decreases especially for the contralateral side and the rear where larger deviations above 5 kHz are observable. Additionally, the reconstructed HRTFs are smoother than the original data for the complete plotted frequency range. If further principal components are considered, they do not further reduce the deviations at higher frequencies. Yet, using more than ten components reduces the remaining error marginally (see Fig. 7.18 or 7.21). In detail, this means that the $ISSD_{dir}$ between measured and reconstructed HRTFs drops below $8\,dB^2$ and only slightly decreases for a larger number of PCs. The remaining deviations exist mainly on contralateral HRTFs.

Complex Spectrum Applying the PCA on the complex spectrum of the HRTFs yields complex-valued principal components which is why they cannot be plotted as in Fig. 7.15. For this reason, the discussion of the influence of the single components and their weights is omitted at this point and instead the focus lies on the number of required components in the following.

---

[6]Since inter-subject differences can be expressed by the scores of the principal components, some studies generate an individualized HRTF data set by subjective tuning of the scores (Shin and Park, 2008; Hwang et al., 2008; Fink and Ray, 2015).

[7]The study by Kistler and Wightman (1992) reported that five components are necessary to reach this threshold, meanwhile the study by Ramos and Tommansini (2014) required six components. Kistler and Wightman (1992) considered 10 subjects, 265 directions and 150 frequency bins, meanwhile Ramos used the data of 47 subjects, 1250 directions and 128 frequency bins. Both studies applied the PCA on the linear magnitudes. For logarithmic magnitudes or a complex spectrum 12 components are necessary to reach the threshold.

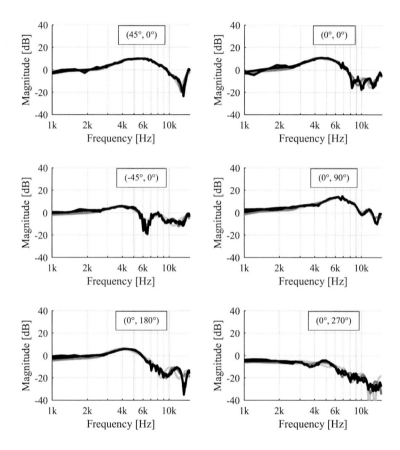

Figure 7.18.: The magnitude spectra of the original HRTFs (black) of data set #17 (randomly chosen) and the reconstructed HRTFs (from light to dark: 6 PCs, 12 PCs and 18 PCs) which were calculated from real-valued PCs, are plotted frequency-dependently for different directions (θ, φ).

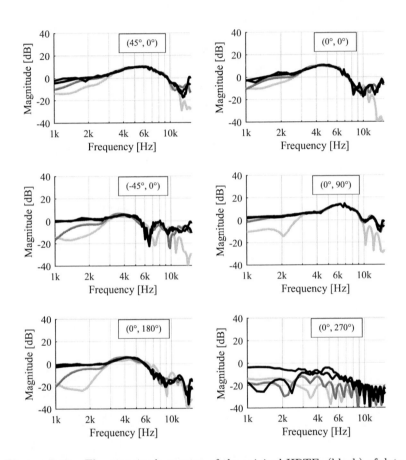

Figure 7.19.: The magnitude spectra of the original HRTFs (black) of data set #17 (randomly chosen) and the reconstructed HRTFs (from light to dark: 6 PCs, 12 PCs and 18 PCs) which were calculated from complex-valued PCs, are plotted frequency-dependently.

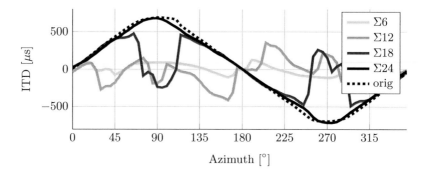

Figure 7.20.: The ITD of data set #17 (randomly chosen) in the horizontal plane was calculated for the measured (black dotted) and reconstructed (gray) HRTFs. The HRTFs were reconstructed from a different number of complex-valued PCs.

The threshold, where 90% of the variances is covered, is reached for 13 components. In contrast to this, the study of Ramos and Tommansini (2014), which considers half as many directions, reached the threshold within the first eight components. The comparison of the reconstructed magnitude spectra in Fig 7.23 shows large deviations when only six or 12 components are considered. These deviations occur mostly below 2 kHz where the reconstructed magnitude decreases in contrast to the measured one, but also for the contralateral HRTFs where 18 components seem not to be sufficient for the reconstruction.

Since the phase is also reconstructed by the complex-valued PCs, the ITD can be determined from the reconstructed HRTF data set. This ITD often deviates strongly from the measured one using less than 21 components (see Fig. 7.20). The deviations occur especially around the interaural axis in the horizontal plane and can be attributed to the low amplitude of the direct arriving sound which is attenuated by the shadowing effect of the head. Due to the fact that this amplitude is only a very small fluctuation compared with the amplitude of the ipsilateral HRIR, it is not considered within the first 21 components. Consequently, the determination of the ITD by the IACC, the THX or the phase delay method fails (cf. Section 5.2).

Considering the ISSD$_{dir}$ as an indicator of the error between the original data and the reconstructed data shows a decreasing error for an increasing number of components in Fig. 7.21. The averaged IESD of 19 dB^2 is reached with seven components (cf. Section 7.3), while this criterion is already reached with the first component in case of real-valued data input for the PCA. If more than 21 components are considered for the reconstruction, the ISSD$_{dir}$ for both variants

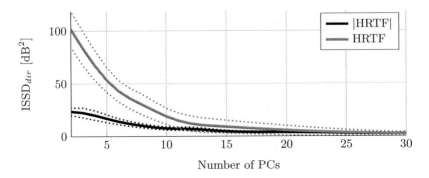

Figure 7.21.: The measure ISSD_{dir} is used to show the difference between the original and the reconstructed data dependent on the considered number of components. The ISSD_{dir}s of the original and reconstructed real-valued data $|\text{HRTF}|$ as well as the ones of the complex-valued data HRTF are shown. The mean is dashed and standard deviation is dotted.

is roughly the same (see Fig. 7.21).

To recap, more than 21 components are necessary to reconstruct the HRTF data set sufficiently by objective criteria such as the magnitude spectrum, the ITD and the ISSD_{dir}. As the reconstruction of the magnitude spectra with a phase replication is more efficient, this method will be further investigated in the following.

Grouping To achieve a better reconstruction with less principal components, multiple PCAs for grouped input data can be applied. For instance, the ipsi- and contralateral HRTFs differ especially in their magnitude which is significantly lower for contralateral HRTFs than for the ipsilateral ones due to the shadowing effect of the head. Therefore each data set was grouped into two data sets: One with ipsilateral HRTFs and the other with the contralateral HRTFs (Braren, 2016). Reconstructing these HRTFs from the two sets of PCs improves the reconstruction when less than 15 real-valued components are considered (see Fig. 7.22). If more than 15 components were used for the reconstruction, the ISSD_{dir} of the grouped and ungrouped versions is comparable. Applying the PCA direction-dependently further reduced error between the reconstructed and measured HRTFs when less than 20 components were considered. Hence, the direction-dependent grouping of the data is more efficient than the ipsi- and contralateral grouping up to 20 components. If more components were considered, there would be no longer be any benefit from grouping.

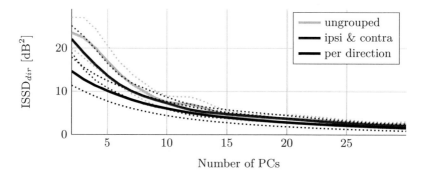

Figure 7.22.: The measure $ISSD_{dir}$ is used to compare grouped PCAs: Ungrouped, ipsi- and contralateral grouped as well as direction-dependent input data. The mean is dashed and standard deviation is dotted.

7.5.2. Anthropometric Estimation of the Spectrum by Principal Components

If individualized HRTF data sets are to be calculated from anthropometric dimensions, the weighting scores $\mathbf{W}$ for the PCs have to be expressed by a linear combination of the anthropometric dimensions (see Section 7.4.2 for further details on the linear regression analysis). Subsequently, the individualized HRTFs can be calculated by these estimated scores $\mathbf{\hat{W}}$ and the PCs (Inoue et al., 2005; Nishino et al., 2007; Hugeng and Gunawan, 2010).

The reconstruction of the HRTFs with the estimated scores works quite well in the median plane where the resonance maximum around 5 kHz and a notch above 6 kHz are observable (cf. upper plots in Fig. 7.23). Meanwhile the magnitude and resonance frequency around 5 kHz are almost aligned with those of the measured data, the first pinna notch does not match perfectly. In general, the measured notch is deeper than the estimated notch. The second maximum of the measured HRTFs around 11 kHz is almost invisible while it is strongly emphasized for the estimated HRTFs. This mismatch between the measured and estimated HRTF at frequencies above 10 kHz is plausible as the wave lengths become very small and the geometry influences the measured HRTFs more strongly.

The increasing mismatch between the estimated and measured HRTFs is also observed for the subject-averaged ISSD in Fig. 7.22. The ISSD rises obviously from 2 to 5 dB around 5 kHz where the spectrum is mainly affected by the notches. Considering more than 18 components does not further reduce the remaining difference between the estimated and measured HRTFs since physical effects at

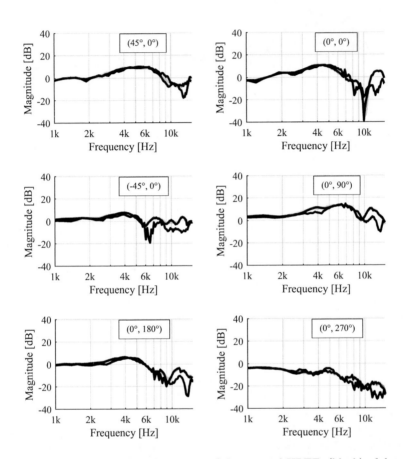

Figure 7.23.: The magnitude spectra of the original HRTFs (black) of data set #17 (randomly chosen) and the anthropometric estimated HRTFs (from light to dark: 6 PCs, 12 PCs and 18 PCs) which were calculated from real-valued PCs, are plotted frequency-dependently for different directions (θ, φ).

the ears and head occur which can no longer be expressed by one-dimensional dimensions. The first nine components already reduce the difference to $16\,\mathrm{dB}^2$ (see Fig. 7.24). If more components are considered, the ISSD_{dir} converges against $15\,\mathrm{dB}^2$ which is comparable to the averaged IESD of $19\,\mathrm{dB}^2$ (cf. Section 7.3). Since the individualization of the HRTF with 13 anthropometric dimensions is demanding, the influence of the dimensions on the accuracy of the reconstructed HRTF is investigated by the regression coefficients $|\beta|$. To find the dimensions with the main impact, they are sorted by their magnitudes $|\beta|$ in an ascending order: h, d_5, d_6, d_1, d_4, d_3, w, d_8, d_2, d_7, d_U, d_M and d_B (dimensions are specified in Fig 4.4). The six dimensions with the largest impact show scores which are twice as large as the other ones. The influence of the head depth for instance is ten times more strongly than the pinna width d_6. Consequently, the benefit of 13 dimensions in comparison to six is roughly $1\,\mathrm{dB}^2$ when more than ten components are considered. Taken together, the reduction of the anthropometric dimensions for the scores from 13 to six does not affect the result greatly. The collection of only six components instead of 13 enables a faster individualization of the HRTF data set with a similar accuracy.

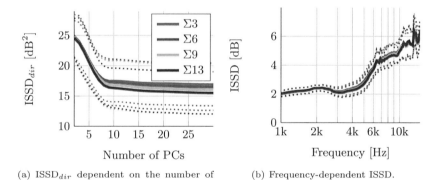

(a) ISSD_{dir} dependent on the number of used components.

(b) Frequency-dependent ISSD.

Figure 7.24.: The spectral difference of the estimated and original HRTF data sets is shown for a different number of anthropometric dimensions. The solid lines mark the mean while dotted lines represent the standard deviation.

7.5.3. Subjective Evaluation of the Individualization by Front-Back Confusions

Evaluating the localization performance of the reconstructed and estimated HRTFs is very challenging and strongly affected by pointing accuracy (see Section 2.1.4). At worst, the pointing inaccuracy may mask the investigated factors. Hence, here only the front-back confusion rate was investigated to show the differences between individual, reconstructed and estimated HRTFs. Reducing the localization task to the decision whether the source is perceived in the frontal or rear hemisphere minimizes the addressed influence of the pointing method on the results. Before discussing the experiment itself, existing subjective studies on the required number of components and the front-back confusion rate are reviewed in the following.

Number of Principal Components Early studies by Kistler and Wightman (1992) as well as Jin et al. (2000) investigated the localization performance under the consideration of one, three and five components[8]. They came to the conclusion that the localization performance in azimuth is mainly provided by the first component and does not improve further by taking more components into account. Contrary to this, the front-back errors dramatically increase for less than five PCs. The same applies to the up-down errors which rise with a decreasing number of components. Both studies reconstructed the HRTFs and did not investigate anthropometric HRTFs for which reason a further experiment was necessary.

Front-Back Confusions In general, the front-back confusion rate depends on the head movements (Wallach, 1940; Makous and Middlebrooks, 1990; Bronkhorst, 1995; Perrett and Noble, 1997; Wightman and Kistler, 1999; Hill et al., 2000), the direction of the incident sound (Kistler and Wightman, 1992; Wenzel et al., 1993), the play-back method (Kistler and Wightman, 1992; Hill et al., 2000) and the listener's experience with the task (Wenzel et al., 1993). The least confusions occur under free-field conditions with free head movements.

The tested spatial source directions vary in terms of number, direction and listener experience for the mentioned studies which is why their confusion rates differ. In contrast to most other above-mentioned studies, the study of Wenzel et al. (1993) tested 16 inexperienced listeners under free-field and headphone conditions at 24 spatially-distributed source positions. For this reason, the results of this study are discussed in the following.

In this study, the confusion rate under free-field conditions reaches $17 \pm 15\%$.

---

[8]Both listening experiments considered five subjects only.

Figure 7.25.: Four parts of the listening experiment.

The confusion rate of sources, which are located in the rear, amounts to $2 \pm 2\%$. Reproducing virtual sources via headphones results in a larger number of confusions. The rate in the front is $25 \pm 15\%$ and drops down to $6 \pm 5\%$ in the rear hemisphere.

While Wenzel et al. (1993) used non-individual HRTFs for the reproduction, Bronkhorst (1995) reported a lower confusion rate using individual HRTFs. Additionally, he showed that stimuli which do not provide spectral cues above $7\,\mathrm{kHz}$ lead to higher confusion rates.

Experience of the Listeners   Usually, inexperienced listeners confuse front-back more often than experienced ones (Wenzel et al., 1993). The averaged confusion rate under free-field conditions was about 7% for experienced and about 32% for inexperienced listeners. The study by Ramos and Tommansini (2014), who tested ten inexperienced listeners with non-individual reconstructed HRTFs, showed a confusion rate of 28% for reconstructed HRTFs from 12 real-valued principal components.

Experimental Setup   Seventeen subjects, on average 25 ± 5 years old, participated in the listening experiment. The individual HRTFs of these subjects had already been measured for the HRTF database (Bomhardt et al., 2016a). Three of these subjects were experienced with binaural reproduction techniques and all others were inexperienced. None of the subjects reported hearing loss or damage.

The experiment consisted of four parts in Fig. 7.25: Reading the instructions, measuring the HpTF of the subjects, performing the test run, and the main experiment.

The HpTFs were measured according to the procedure described in Section 2.6 with Sennheiser HD 650 headphones and KE3 microphones. The measured HpTFs were multiplied with the HRTFs and convolved with a triple pulsed noise stimulus. Each pink noise pulse of $150\,\mathrm{ms}$ was followed by a break of $150\,\mathrm{ms}$. Due to the right ear advantage only directions on the right ear side in the horizontal plane were chosen (Macpherson and Middlebrooks, 2002): Two at the front ($-10°$

and $-30°$), two at the side ($-60°$ and $-70°$) and two at the rear ($-135°$ and $-165°$).

The main experiment consisted of five permuted blocks with different types of HRTFs:

Individual: Individual HRTFs from the database (Bomhardt et al., 2016a).

Symmetric: Symmetric individual HRTFs (the individual HRTFs were mirrored on the median plane).

$PC \in \mathbb{R}$: Reconstructed individual HRTFs from 15 real-valued PCs with optimal weighting scores (the ITD was estimated by (5.19) with an optimized radius from the present database).

$PC_a^+ \in \mathbb{R}$: Estimated symmetric HRTFs from 15 real-valued PCs with anthropometric weighting scores (the ITD was estimated by (5.19) with an optimized radius, and the PCA was applied on the complete database).

$PC_a^- \in \mathbb{R}$: Estimated symmetric HRTFs from 15 real-valued PCs with anthropometric weighting scores (the ITD was estimated by (5.19) with an optimized radius, and the PCA did not consider the HRTF data set of the tested subject).

Each block was started by the subject and began with the play-back of an arbitrary chosen stimulus. The directions of the stimulus were tested randomly five times per block. After the play-back, the subject had to choose one of the buttons according to the following instructions:

Front: If the subject perceived the sound from the frontal quadrant, the subject was to choose the button *Front*.

Rear: If the subject perceived the sound in the rear quadrant, the subject was to choose the button *Rear*.

Confusion: In case the subject had an in-head or an ambiguously perceived direction, the subject was to choose *Confusion*.

After 30 trials the block ended and the next block was started 30 seconds later. In the trial run three directions of a foreign HRTF data set were chosen to prepare the subjects for the main task. Apart from this, the procedure of the trial run coincided with that of the main experiment.

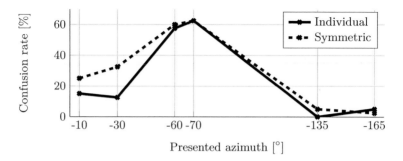

Figure 7.26.: The front-back confusion rate over the azimuth angle for individual and symmetric HRTFs. The dots mark the directions presented.

The whole experiment took about 30 to 40 minutes.

Evaluation First, the results of the control blocks with the individual and symmetric HRTFs are discussed; subsequently, those of the reconstructed and estimated HRTFs are considered.

To evaluate the result of this listening test, the number of front-back and back-front confusions was added and divided by the total number of trials per direction. In the following, there is no distinction between front-back confusions and back-front confusions. Both confusion types are summarized as front-back confusion. The average front-back confusion rate of the subjects[9] with individually measured HRTFs is 14% at the front, rises to 59% for lateral directions and drops again to 3% at the rear (see Fig. 7.26). At the front the rate is the lowest for the individual HRTF data sets and at the sides the error is relatively large in both cases. An explanation for the large lateral error might be that the subjects defined their interaural axis, which splits the frontal and rear quadrant, in front of the ears. Consequently, these virtual sources close to the interaural axis were often rated as sources at the rear. Apart from the lateral directions, the rates are in agreement with rates in the study by Wenzel et al. (1993). As previously mentioned, a comparison of the confusion rates between studies is difficult since the rate depends on the tested directions, experience of the listeners and the play-back method.

Comparing the confusion rates of the individual and symmetric HRTFs, lateral and rear directions are very similar. The frontal rate with the symmetric HRTFs exceeds those of the individual HRTF. It is assumed that this difference can be

---

[9]Due to a technical error in the experiment for the first nine subjects in the block with the symmetric HRTFs, the results of only eight subjects remained.

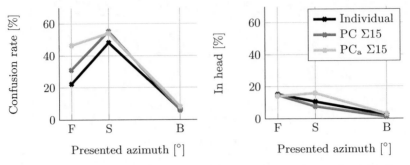

(a) Azimuth-dependent front-back confusion rate.

(b) Azimuth-dependent in head localizations.

Figure 7.27.: The responses of the subjects were evaluated for individual, optimal and anthropometric reconstructed HRTFs. The dots mark the presented directions $F \in \{-30°, -10°\}$, $S \in \{-70°, -60°\}$ and $B \in \{-165°, -135°\}$.

attributed to the missing asymmetry of the individual HRTF.

Evaluating the subject-dependent deviations for each direction results in large deviations due to the few repetitions[10]. For that reason, a statistical analysis showed no significance between the HRTF types. Furthermore, Fig. 7.26 shows that the confusion rates do no vary strongly between frontal, lateral and rear directions. For this reason they will be grouped in the following ($F \in \{-30°, -10°\}$, $S \in \{-70°, -60°\}$ and $B \in \{-165°, -135°\}$).

The measured, reconstructed and estimated HRTFs are compared in the following. For this purpose the front-back confusions of the individual, reconstructed and anthropometric estimated HRTFs[11] are plotted in Fig. 7.27. The default confusion rate with individual HRTFs is 22% at the front, 48% at the side and 6% at the rear. The results of the reconstructed HRTFs deviate particularly for frontal directions and only slightly for lateral ones[12]. The rates increase from 22% for individual to 31% for reconstructed and 46% for the anthropometric estimated HRTFs at the front. At the rear the rates are very similar around 7% for all three types. The comparison of the rates for lateral directions is difficult due to the already mentioned issues with the perceived interaural axis.

---

[10] Repeating each direction five times per subject results in discrete steps of 20% which is very imprecise for a statistical subject-dependent evaluation.

[11] For the evaluation of these blocks, the results of all 17 subjects are considered.

[12] The results of the 17 subjects show the same tendencies as the ones of the eight subjects in Fig. 7.26. The slightly higher rate of confusion in the front and the lower rate for lateral directions can be attributed to the larger number of subjects. In general, all frontal, lateral and rear directions show similar confusion rates. By this reason, they are grouped.

The in-head localizations decrease for the individual and reconstructed HRTFs towards the back from 14% to 2%. The number of in-head localizations for lateral sources with anthropometric estimated HRTFs is twice as large as the one with the reconstructed HRTF data sets. Due to the fact that the reconstruction does not show the same tendency, it has to be an effect of insufficiently estimated HRTFs. Hence, an improvement of the individualization is desirable for these directions.

The results of the blocks with the anthropometric estimated HRTFs ($PC_a^+ \in \mathbb{R}$ and $PC_a^- \in \mathbb{R}$) do not differ remarkably. Thus, it is not necessary to exclude the individual HRTF data set of the tested subject for the PCA.

In order to find possible influencing factors for the higher confusion rates of the symmetric, reconstructed and estimated HRTFs, the HRTFs of four particular subjects (#6, #35, #31 and #46) were compared informally. The symmetric and estimated HRTFs showed differences especially at frequencies above 6 kHz where smaller dimensions affect the HRTFs and the estimated HRTF underrates the notch. However, no clear difference could be observed and attributed to the confusion rate. In order to clarify these relationships between confusion rate and HRTF deviations, a comprehensive listening experiment with experts and non-experts should be carried out in future.

To conclude, the number of front-back confusions rises for symmetric, reconstructed and estimated HRTFs at the front and is otherwise almost similar to those of the individual HRTF. Especially symmetric and smooth estimated HRTFs seem to impair the localization. The performance with reconstructed HRTFs from 15 real-valued components with an estimated phase is only slightly worse than the one with individual HRTFs.

7.6. Comparison of the Methods

This final section compares the different individualization approaches introduced in this chapter. For this purpose, the optimally scaled HRTFs are contrasted with the reconstructed HRTFs from the PCA and the anthropometrically scaled HRTFs with the estimated HRTFs from the anthropometric scores.

The accuracy of these approaches is evaluated by the following measures:

$ISSD_{dir}$: The inter-subject spectral difference compares the individually measured HRTF data set with those to be tested[13]. The difference was calculated in the frequency range between 1 and 13 kHz for all HRTF data sets of the present database.

---

[13] The HRTFs to be tested are the scaled, reconstructed and anthropometrically estimated HRTFs.

Δf_P: The first resonance frequency of the HRTF data sets was determined from the direction-averaged HRTF of each subject. The resonance frequency mismatch Δf_P between the measured and tested HRTF data sets was calculated by the difference of both.

ΔA_P: The direction-averaged resonance maximum was determined from the direction-averaged HRTF of each subject. The resonance magnitude error ΔA_P between the measured and tested HRTF data sets was calculated by the difference of both.

Δf_N: Since the notches of the HRTF data sets are clearly apparent in the frontal hemisphere, the proposed detection strategy from Section 7.1.2 was used to determine the notch frequencies. Based on the fact that the notch occurs direction-dependently, it was determined for the right ear at $(\theta, \varphi) = (-30°, 300°)$. The notch frequency error Δf_N between the measured and tested HRTF data sets was calculated by the difference of both.

ρ_P: In general, it is assumed that the first resonance frequencies of the measured HRTF data set and HRTF data set to be tested are linearly connected. The correlation coefficient of both provides further information on their deviations.

ρ_A: The correlation coefficient of the detected resonance magnitudes of two HRTF data sets was calculated to show deviations between both.

ρ_N: The correlation coefficient of the detected notch frequencies of two HRTF data sets was calculated to show deviations between both.

All listed measures, excluding the ISSD_{dir}, are related to the physical influence of the human body on the HRTF (cf. Section 7.1). While the characteristic of the first resonance is more important for the ILD, the notch provides essential information for the localization in elevation and the front-back discrimination (cf. Section 2.1). The phase or ITD is not evaluated in this section due to the fact that the phase of the real-valued PCA has to be estimated by a minimum phase and an all-pass. The delay of this all-pass is dependent on the ITD model used. The accuracy of the estimated phase (minimum phase and ITD) can be studied in Chapter 5. If a complex-valued PCA is performed, the accuracy of the phase is shown in Fig. 7.20.

Before considering the performance of the estimated HRTF data sets by these measures, the compared HRTF data sets are summarized:

s: The HRTF data sets of the present database were used to obtain the optimal scaling factor s for the data set of subject #19. This data set is one of the best matching data sets of the present database. It shows a low averaged $ISSD_{dir}$ of $28\,dB^2$ with regard to the other data sets (cf. Section 7.4.1).

$PC \in \mathbb{R}$:   The PCA was applied on the whole present database with the linear magnitudes as input data. In total, 15 principal components were used for the reconstruction.

$PC \in \mathbb{C}$:   Instead of taking the linear magnitudes, the complex-valued PCA used the complex spectrum of the HRTF data sets as data input. Consequently the resulting principal components and their scores are complex-valued. According to the PCA with real-valued data input, 15 complex-valued components were used for the reconstruction. The reconstruction with 21 complex-valued components will lead to a comparable magnitude to the real-valued components.

$\hat{s}$:   If a given the HRTF data set should be individualized by the geometry of a subject, the relationship between the optimal scaling factors of this data set and the anthropometric dimensions has to be determined first (cf. Section 7.4.2). Subsequently, this data set can be scaled by the anthropometric dimensions of arbitrary subjects. In accordance with the optimal scaling factor, the well-matching data set of subject #19 was used for the scaling with the anthropometrically estimated factor.

$PC_a \in \mathbb{R}$:   The principal components can also be used to estimate an HRTF data set by the anthropometric dimensions of an arbitrary subject. Since more than 15 components do not significantly improve the estimated HRTF data sets, this number of real-valued components was used here.

The accuracy of these reconstruction and estimation approaches compared to individual HRTFs can be found in Tab. 7.8.

The subject-averaged $ISSD_{dir}$ of each approaches has already been discussed separately in the previous sections. In the present comparison, the real-valued reconstruction with 15 components outperforms the reconstruction with complex-

	s	$PC \in \mathbb{R}$	$PC \in \mathbb{C}$	$\hat{s}$	$PC_a \in \mathbb{R}$
ISSD [dB^2]	24 ± 5	5 ± 1	9 ± 6	26 ± 5	17 ± 4
Δf_P [kHz]	0.0 ± 0.4	0.0 ± 0.1	0.1 ± 0.3	0.0 ± 0.5	0.1 ± 0.5
ΔA_P [dB]	-2.1 ± 1.7	0.0 ± 0.1	0.2 ± 0.3	-2.1 ± 1.7	0.1 ± 1.5
Δf_N [kHz]	0.3 ± 0.3	0.0 ± 0.3	0.0 ± 0.2	0.3 ± 0.4	0.1 ± 0.4
ρ_P	0.55	0.99	0.77	0.29	0.34
ρ_A	-0.12	1.00	0.98	-0.03	0.45
ρ_N	0.81	0.86	0.92	0.69	0.62

Table 7.8.: Different reconstruction and estimation approaches for HRTF data sets are compared with the individual HRTF data sets in this table. For this purpose, the ISSD$_{dir}$, the difference in the resonance frequency and its maximum as well the notch are used. For all measures the mean over all subjects and its standard deviation are shown.

valued components as well as the optimal scaling. The same applies for the anthropometric approaches where the ISSD$_{dir}$ of the real-valued components is smaller than that of scaled data sets.

Comparing the error of the first resonance frequencies, it is obvious that the error of the real-valued components is smaller than that of all other approaches. In contrast to this, the magnitude of the scaled data set varies more strongly due to the fact that scaling does not adapt the magnitude itself. As already shown in Fig. 7.21, the accordance with the resonance frequency of the complex-valued PCA is worse than for the real-valued PCA. While the reconstructed data sets fit well with the measured ones, the data sets with the anthropometric scores show deviations from the measured ones which are larger than 1 dB.

Since outliers are not detectable by the mean and standard deviation, the correlation of the resonance frequencies of measured and scaled data sets show that these frequencies do not always fit very well. For the reconstruction techniques these correlations are very high $|\rho_P| > 0.7$ compared to the scaling. As far as anthropometric approaches are concerned, the correlation coefficient drops down below 0.4. One explanation for this is that the half width of the resonance is very large so that small measurement uncertainties influence the determination of the resonance frequency (cf. Section 7.1.1). The same applies to its magnitude. The adaption of the magnitudes works well for the reconstruction techniques while the estimation by anthropometric dimensions shows only a moderate correlation $0.3 \leq |\rho_A| \leq 0.5$. Anyway, the scaling allows no possibility to adapt the magnitude.

In general, the notch frequency for the direction $(\theta, \varphi) = (-30°, 300°)$ fluctuates for the considered approaches. While variations of $0.3\,\mathrm{kHz}$ can be expected for the optimal scaling and the reconstruction, the mismatch for the anthropometric data sets can be larger than $0.4\,\mathrm{kHz}$. In this case, the averaged error of the anthropometrically scaled data set is larger than the one of the data set with the anthropometric scores. One reason for this is that the notch of the anthropometric data sets is often relatively weak and therefore more difficult to determine. However, the correlation coefficient of the anthropometric scaling is also larger than the one of the optimal scaling which indicates that the notch frequency estimation approach is very sensitive to measurement uncertainties. It should be noted that the first notches of the anthropometric reconstructed data sets are less pronounced than in case of scaling.

To conclude, the reconstruction with 15 components outperforms the optimal scaling. If the HRTF data set is to be estimated, the data sets with the anthropometric scores match better than the anthropometrically scaled ones. However, the notch frequency, which is important for the front-back discrimination, is slightly better estimated by scaling.

8

Conclusion and Outlook

This thesis reveals the link between individual HRTFs and anthropometric dimensions with the objective to adapt an HRTF data set using the subject's head and pinna dimensions.

For this purpose, a database containing the HRTF data sets of 48 subjects and their ear models as well as head dimensions were measured (Chapter 4).

Since the interaural time difference (ITD) mainly contributes to the sound source localization at low frequencies, different methods for the determination of the ITD in time- and frequency-domains were discussed in Chapter 5. Subsequently, objective and subjective measures have been developed to compare different ITDs from the presented database.

While humans use the ITD at low frequencies to localize sound sources, spectral cues such as interference effects or interaural level differences provide spatial information for the auditory system at higher frequencies. The interference effects are either characterized by maxima or notches of the HRTFs. Since these spectral cues are important to localize sound sources, approaches to detect these effects have been introduced and later used to evaluate the quality of the individualization at higher frequencies.

Two existing approaches to adapt the spectrum of the HRTF were studied, improved and compared in Chapter 7: The frequency scaling (Middlebrooks, 1999b,a) and the decomposition into principal components (Ramos and Tommansini, 2014; Nishino et al., 2007).

The first approach improved the localization performance of a listener by scaling or squeezing the frequency vector of a non-individual HRTF data set. The optimal scaling factor was obtained from the overall spectral difference between the data set to be scaled and the data set of the subject which was to be adapted. The scaling of the non-individual HRTF data set revealed that the spectral cues of this set fitted better to the one of the listener than a non-individual set. The quality of this fit depends on the selected data set to be adapted: Some of the sets matched well for most of the subjects while others deviated substantially. Using the link between the spectral cues and the anthropometric dimensions, the scaling factor could be expressed by the dimensions of the head and ear. The accuracy of the anthropometrically scaled HRTF data sets was slightly worse than the optimally scaled sets but better than the one of non-individual sets. Especially narrow band notches of the adapted HRTF corresponded well to individual ones. Due to the squeezing of the frequency vector, the adaption of the magnitude was not feasible. In addition, this approach squeezed the phase and therefore the ITD which could lead to a mismatched ITD.

The second approach decomposed the HRTF data sets into principal components and direction- and subject-dependent weighting scores. Applying the PCA on the complex-valued spectrum, the phase and the spectral cues could be reconstructed. In this case, more than 20 principal components were necessary to reconstruct contralateral HRTFs with an adequate ITD. When real-valued magnitudes were used as data input for the PCA, significantly fewer components were necessary. However, the phase cannot be reconstructed from real-valued components and weighting scores. While the accuracy of the reconstructed HRTF data sets improved with a higher number of considered components, the error for the anthropometrically estimated weighting scores converged towards $16\,dB^2$. This can be explained by the accuracy of small one-dimensional measures which influence especially higher frequencies. Therefore, a significant reduction of the error was possible using 8 PCs. More than 10 PCs did not significantly improve the estimated HRTF data set. However, the first resonance frequency and magnitude were very well estimated, whereas the first notch of the spectrum was often not modeled adequately. Reducing the number of anthropometric dimensions from 13 to 6 did not significantly influence the adaption. This reduction has the advantage that fewer anthropometric dimensions have to be measured.

Using real-valued components to reconstruct or estimate an HRTF data set, the phase has to be estimated by a minimum phase plus the ITD. Since the minimum phase can be calculated from the magnitude spectrum, the ITD had to be modeled by anthropometric dimensions. Two different models are proposed in this thesis (see Chapter 5). The first one is an analytic model based on the geometry of an ellipsoid. This model shows comparable results to already existing models (Woodworth, 1940; Kuhn, 1977; Larcher and Jot, 1997; Savioja et al., 1999). The second one is an empirical model on the basis of two anthropometric dimensions. This model adapts the ITD better than the ellipsoid but still shows deviations to the measured data due to the missing asymmetry and the approximated curve progression.

Subjective investigations on the audible ITD error by a listening experiment revealed that the error for lateral directions can be larger than for frontal ones. Consequently, the subjects were more sensitive to a mismatching ITD for frontal and oblique directions than for lateral ones. Comparing the just noticeable ITD error with the error of the proposed ITD models showed that the modeled anthropometric ITDs fitted very well.

The subjective evaluation of the individualized HRTF data sets was performed by a listening experiment. In this experiment the front-back confusions, which can be solved by monaural cues and head movements, were investigated. Anthropometrically estimated HRTFs (PCA) and individual HRTFs showed a similar confusion rate for rear directions. Regarding frontal directions, the individual HRTFs outperformed the estimated HRTFs due to missing spectral cues. However, in this study, differences between the reconstructed HRTFs and individual HRTFs were still observable for frontal directions. The frequency scaling was not investigated in this experiment since Middlebrooks (1999b) had already shown that lower $ISSD_{dir}$ between two HRTF data sets improves the front-back confusion rate.

Furthermore, the symmetry of the ear has been investigated since the assumption of symmetric ears and HRTFs reduces the measurement effort for the anthropometric dimensions. For this purpose, the one-dimensional pinna dimensions of both ears were compared. Since these dimensions deviated by only 1 to 2 mm, it was assumed that the ears are almost symmetric. The same applied to the mismatch between HRTFs of the right and left ears.

To summarize, the estimation of the anthropometric HRTF was feasible by scaling or principal components. The individualized HRTFs using real-valued principal components were slightly better than frequency scaling since the magnitude and the frequency can be adapted. However, the narrow band notches, which are important for front-back discrimination and localization in elevation, were more distinct by frequency scaling. The estimation of the phase did not represent a problem, since it was feasible to estimate the ITD by an adequate ITD model (Chapter 5).

In conclusion, the presented anthropometric individualization of HRTFs enhances the localization performance in virtual environments without troublesome and time-consuming acoustic measurements.

A

Kalman Filter for Minima Detection

In Section 7.1.2 a local minima detection in combination with a Kalman filter is used to determine the notches of an HRTF data set by the following steps (Bomhardt, 2016; Bomhardt and Fels, 2017):

1. The ipsilateral HRTFs are extracted from an HRTF data set for a single azimuth angle φ, since the notch is less pronounced for contralateral HRTFs.

2. A local minimum detection is applied on the selected transfer functions between 3 kHz and 11 kHz under the assumption that the minimum has to be smaller than 5 dB. To detect these minima by a local maximum search, the magnitudes of the transfer functions are inverted. Previously, the magnitudes were lowered by 20 dB to avoid zero divisions. The local maxima search is limited on magnitudes which are greater than $-\frac{1}{15}$ $^1/_{\mathrm{dB}}$ and drop at least 10^{-5} $^1/_{\mathrm{dB}}$.

3. A Kalman filter (Bishop and Welch, 2001) is exploited as a tracking algorithm for the first notch with an underlying state transition model. Measurement noise and disturbances of the tracked notch can be considered by this filter so that the estimation is very robust.

   Since the notch is shifted elevation-dependent to higher frequencies on a logarithmic scale (see Fig. 7.6), the logarithmic frequency is considered in the following. In most of the HRTF data sets, the starting point can be well determined around $\theta_0 = -60°$ and $f_0 = 6\,\mathrm{Hz}$. Based on this fact, an initial position $\mathbf{x}_0$ is chosen

$$\mathbf{x}_0 \quad = \quad [\theta_0 \quad \log_{10}(f_0)]'. \tag{A.1}$$

Considering that the position of the logarithmic notch frequency f_k rises linearly with an increasing elevation angle θ_k, the next position $\mathbf{x}_{k+1}$ can be estimated by the previous one $\mathbf{x}_k$. Consequently, the resulting state transition model is defined as

$$\mathbf{A} = \begin{bmatrix} 1 & 1 \\ 0 & 1 \end{bmatrix} \tag{A.2}$$

so that the estimation of the next position $\mathbf{x}_{k+1}$ can be calculated as

$$\mathbf{x}_{k+1} \quad = \quad \mathbf{A} \cdot \mathbf{x}_k + \mathbf{w}_k. \tag{A.3}$$

A process noise $\mathbf{w}_k$, which is derived from the covariance matrix $\mathbf{Q} = \operatorname{diag}(\theta_{motion}, \log_{10}(f_{motion}))$, is considered in this equation to cover the uncertainties of the estimation (so-called motion noise).

The measured position $\mathbf{z}_k$, which can be found by a nearest neighbor search from the detected minima, is described in dependency of the estimated position $\mathbf{x}_k$

$$\mathbf{z}_k \quad = \quad \mathbf{H} \cdot \mathbf{x}_k + \mathbf{v}_k. \tag{A.4}$$

The measurement model $\mathbf{H}$ takes only the estimated position into account wherefore it is defined as $\mathbf{H} = [1 \quad 0]$. Furthermore, the observed position $\mathbf{z}_k$ is affected by measurement noise $\mathbf{v}_k$ which is considered by the measurement noise covariance matrix $\mathbf{R} = \operatorname{diag}(\theta_{meas}, \log_{10}(f_{meas}))$.

In a final step, the Kalman filter is updated using a gain factor K which is given by the estimated accuracy of the state estimate $\mathbf{P} = \operatorname{diag}(\theta_{err,ini}, \log_{10}(f_{err,ini}))$, the measurement model $\mathbf{H}$, and measurement noise covariance matrix $\mathbf{R}$. By the help of this gain factor K, the updated state estimate (position) and estimated covariance matrix is determined for the next step[1].

4. To interpolate the notch positions found, the detected notches are fitted by the linear approximation $\theta_N = m \cdot \log_{10}(f) + n$.

---

[1] In the implementation of this thesis, the logarithmic notch frequency is weighted by an additional factor which is not considered in the current equations. Based on this fact, the subjectively chosen initial error $\mathbf{P}$, process noise $\mathbf{Q}$ and measurement noise covariance matrix $\mathbf{R}$ are not are not further specified here.

B

Linear Regression Analysis

In this thesis often a linear relationship between anthropometric dimensions α_i and an HRTFs feature ζ is assumed. Based on this relationship, this feature $\hat{\zeta}$ can be approximated by weighted anthropometric dimensions (cf. Section 7.4.2 or 7.5.2).

In total, n_{anthro} anthropometric dimensions are given by the database at hand for each subject #j. Consequently, the feature ζ can be approximated by a linear combination of anthropometric features $\alpha_{j,i}$ and regression coefficients β_i (Seber, G. A. F and Lee, 2003, pp. 1-12)

$$\hat{\zeta}_j = \beta_0 + \sum_{i=1}^{n_{anthro}} \beta_i \, \alpha_{j,i}, \tag{B.1}$$

$$= \boldsymbol{\alpha}_j \cdot \boldsymbol{\beta}. \tag{B.2}$$

The anthropometric dimensions are summarized by $\boldsymbol{\alpha}_j = [1 \; \alpha_{j,1} \ldots \alpha_{j,n_{anthro}}]$ and the coefficients by $\boldsymbol{\beta} = [\beta_0 \ldots \beta_{n_{anthro}}]'$. Since only a finite number n_{anthro} of anthropometric features can be considered and a rough linear approximation is used, an error term

$$e_j = \zeta_j - \hat{\zeta}_j \tag{B.3}$$

remains. To derive the estimated regression coefficients $\hat{\boldsymbol{\beta}}$, the quadratic error term of the vector $\mathbf{e} = [e_1 \ldots e_{n_{subj}}]'$ of n_{subj} subjects has to be minimized

$$\hat{\boldsymbol{\beta}} = \arg\min_{\boldsymbol{\beta}} \sum_{j=1}^{n_{subj}} |e_j|^2, \tag{B.4}$$

$$= (\mathbf{A}\,\mathbf{A}')^{-1}\,\mathbf{A}\,\boldsymbol{\zeta}. \tag{B.5}$$

Hereby, the anthropometric dimensions are given by $\mathbf{A} = [\boldsymbol{\alpha}'_1 \ldots \boldsymbol{\alpha}'_{n_{subj}}]'$ and the HRTF feature by $\boldsymbol{\zeta} = [\zeta_1 \ldots \zeta_{n_{subj}}]'$. Subsequently, the estimated anthropometric scaling factors $\hat{\zeta}$ can be expressed as

$$\hat{\zeta} = \mathbf{A}' \, \hat{\beta}. \tag{B.6}$$

Bibliography

Abdi, H. and Williams, L. J. (2010). Principal component analysis. *Wiley Interdisciplinary Reviews: Computational Statistics*, 2(4):433–459.

Algazi, V. R., Avendano, C., and Duda, R. O. (2001a). Estimation of a spherical-head model from anthropometry. *Journal of the Audio Engineering Society*, 49(6):472–479.

Algazi, V. R., Duda, R. O., Duraiswami, R., Gumerov, N. A., and Tang, Z. (2002a). Approximating the head-related transfer function using simple geometric models of the head and torso. *The Journal of the Acoustical Society of America*, 112(5):2053–2064.

Algazi, V. R., Duda, R. O., Morrison, R. P., and Thompson, D. M. (2001b). Structural composition and decomposition of HRTFs. In IEEE, editor, *2001 IEEE Workshop on the Applications of Signal Processing to Audio and Acoustics*.

Algazi, V. R., Duda, R. O., and Thompson, D. M. (2001c). The CIPIC HRTF database. In IEEE, editor, *2001 IEEE Workshop on the Applications of Signal Processing to Audio and Acoustics*.

Algazi, V. R., Duda, R. O., and Thompson, D. M. (2002b). The use of head-and-torso models for improved spatial sound synthesis. In Audio Engineering Society, editor, *113rd Audio Engineering Society Convention*.

Andreopoulou, A., Begault, D. R., and Katz, B. F. G. (2015). Inter-laboratory round robin HRTF measurement comparison. *IEEE Journal of Selected Topics in Signal Processing*, 9(5):895–906.

Andreopoulou, A., Roginska, A., and Mohanraj, H. (2013). A database of repeated head-related transfer function measurements. In International Community for Auditory Display, editor, *International Conference on Auditory Display*.

Aussal, M., Algouges, F., and Katz, B. F. G. (2012). ITD interpolation and personalization for binaural synthesis using spherical harmonics. In Audio Engineering Society, editor, *25th AES UK Conference: Spatial Audio in Today's 3D World*.

Bahu, H., Carpentier, T., Noisternig, M., and Warusfel, O. (2016). Comparison of Different Egocentric Pointing Methods for 3D Sound Localization Experiments. *Acta Acustica united with Acustica*, 102(1):107–118.

Bishop, G. and Welch, G. (2001). An introduction to the kalman filter. *Proc of SIGGRAPH, Course*, 8(27599-23175):41.

Blauert, J. (1969). Sound localization in the median plane. *Acta Acustica united with Acustica*, 22(4):205–213.

Blauert, J. (1997). *Spatial hearing: the psychophysics of human sound localization.* MIT press, Cambridge, Massachusetts and London, England, 2nd edition.

Blauert, J., Brüggen, M., Hartung, K., Bronkhorst, A. W., Drullmarm, R., Reynaud, R., Pellieux, L., Krebber, W., and Sottek, R. (1998). The AUDIS catalog of human HRTFs. In Acoustical Society of America, editor, *16th International Congress on Acoustics and the 135th Meeting of the Acoustical Society of America.*

Bloom, P. J. (1977). Creating source elevation illusions by spectral manipulation. *Journal of the Audio Engineering Society,* 25(9):560–565.

Bomhardt, R. (2016). Detection of Notches in Head-Related Transfer Functions. In CTU Prague, editor, *20th International Student Conference on Electrical Engineering.*

Bomhardt, R., de la Fuente Klein, M., and Fels, J. (2016a). A high-resolution head-related transfer function and three-dimensional ear model database. In Acoustical Society of America, editor, *5th Joint Meeting of the Acoustical Society of America and Acoustical Society of Japan.*

Bomhardt, R. and Fels, J. (2014). Analytical interaural time difference model for the individualization of arbitrary Head-Related Impulse Responses. In Audio Engineering Society, editor, *137th International AES Convention.*

Bomhardt, R. and Fels, J. (2016). Mismatch between Interaural Level Differences Derived from Human Heads and Spherical Models. In Audio Engineering Society, editor, *140th International AES Convention.*

Bomhardt, R. and Fels, J. (2017). The influence of symmetrical human ears on the front-back confusion. In Audio Engineering Society, editor, *142nd Audio Engineering Society Convention.*

Bomhardt, R., Lins, M., and Fels, J. (2016b). Analytical Ellipsoidal Model of Interaural Time Differences for the Individualization of Head-Related Impulse Responses. *Journal of the Audio Engineering Society,* 64(11):882–894.

Braren, H. (09.05.2016). Analysis and Morphing of Monaural Cues from Individual Head-Related Transfer Functions. Master's thesis.

Breebaart, J. and Kohlrausch, A. (2001). The perceptual (ir) relevance of HRTF magnitude and phase spectra. In Audio Engineering Society, editor, *110th Audio Engineering Society Convention.*

Bronkhorst, A. W. (1995). Contribution of spectral cues to human sound localization. *The Journal of the Acoustical Society of America,* 98(5):2542–2553.

Bronstein, I. N. and Semendjajew, K. A. (1991). *Taschenbuch der Mathematik.* Harri Deutsch, Frankfurt/Main, Germany, 25th edition.

Brungart, D. S. and Rabinowitz, W. M. (1999). Auditory localization of nearby sources. Head-related transfer functions. *The Journal of the Acoustical Society of America,* 106(3):1465–1479.

Busson, S. (01.01.2006). *Individualisation d'indices acoustiques pour la synthèse binaurale*. PhD thesis, Université de la Méditerranée-Aix-Marseille II.

Butler, R. A. and Belendiuk, K. (1977). Spectral cues utilized in the localization of sound in the median sagittal plane. *The Journal of the Acoustical Society of America*, 61:1264.

Carlile, S. and Pralong, D. (1994). The location–dependent nature of perceptually salient features of the human head–related transfer functions. *The Journal of the Acoustical Society of America*, 95:3445.

Chan, J. C. K. and Geisler, C. D. (1990). Estimation of eardrum acoustic pressure and of ear canal length from remote points in the canal. *The Journal of the Acoustical Society of America*, 87(3):1237–1247.

Chen, J., van Veen, B. D., and Hecox, K. E. (1995). A spatial feature extraction and regularization model for the head-related transfer function. *The Journal of the Acoustical Society of America*, 97:1.

Damaske, P. and Wagener, B. (1969). Investigations of directional hearing using a dummy head. *Acustica*, 21:30–35.

Dellepiane, M., Pietroni, N., Tsingos, N., Asselot, M., and Scopigno, R. (2008). Reconstructing head models from photographs for individualized 3D–audio processing. *Computer Graphics Forum*, 27:7.

Deschrijver, D., Haegeman, B., and Dhaene, T. (2007). Orthonormal vector fitting: A robust macromodeling tool for rational approximation of frequency domain responses. *IEEE Transactions on advanced packaging*, 30(2):216.

Dietrich, P. (2013). *Uncertainties in Acoustical Transfer Functions: Modeling, Measurement and Derivation of Parameters for Airborne and Structure-borne Sound*. Dissertation, RWTH Aachen University, Aachen.

Driscoll, J. R. and Healy, D. M. (1994). Computing Fourier transforms and convolutions on the 2-sphere. *Advances in applied mathematics*, 15(2):202–250.

Duda, R. O. and Algazi, V. R. (1999). An adaptable ellipsoidal head model for the interaural time difference. In IEEE, editor, *1999 IEEE International Conference on Acoustics, Speech, and Signal Processing*.

Duraiswami, R., Zotkin, D. N., and Gumerov, N. A. (2004). Interpolation and range extrapolation of HRTFs. In IEEE, editor, *2004 IEEE International Conference on Acoustics, Speech, and Signal Processing*.

Emmerich, D. S., Harris, J., Brown, W. S., and Springer, S. P. (1988). The relationship between auditory sensitivity and ear asymmetry on a dichotic listening task. *Neuropsychologia*, 26(1):133–143.

Fahy, F. J. (1995). The vibro-acoustic reciprocity principle and applications to noise control. *Acta Acustica united with Acustica*, 81(6):544–558.

Fels, J., Buthmann, P., and Vorländer, M. (2004). Head-related transfer functions of children. *Acta Acustica united with Acustica*, 90(5):918–927.

Fels, J. and Vorländer, M. (2009). Anthropometric parameters influencing head-related transfer functions. *Acta Acustica united with Acustica*, 95(2):331–342.

Fink, K. J. and Ray, L. (2015). Individualization of head related transfer functions using principal component analysis. *Applied Acoustics*, 87:162–173.

Gardner, B. and Martin, K. (1994). HRFT Measurements of a KEMAR Dummy-head Microphone. *MIT Media Lab Perceptual Computing*.

Giguère, C. and Abel, S. M. (1993). Sound localization: Effects of reverberation time, speaker array, stimulus frequency, and stimulus rise/decay. *The Journal of the Acoustical Society of America*, 94(2):769–776.

Guillon, P., Guignard, T., and Rozenn, N. (2008). Head-Related Transfer Function customization by frequency scaling and rotation shift based on a new morphological matching method. In Audio Engineering Society, editor, *125th AES Convention*.

Gumerov, N. A., O'Donovan, A. E., Duraiswami, R., and Zotkin, D. N. (2010). Computation of the head-related transfer function via the fast multipole accelerated boundary element method and its spherical harmonic representation. *The Journal of the Acoustical Society of America*, 127(1):370–386.

Gupta, N., Barreto, A., Joshi, M., and Agudelo, J. C. (2010). HRTF Database at FIU DSP Lab. In IEEE, editor, *2010 IEEE International Conference on Acoustics Speech and Signal Processing*.

Gustavsen, B. and Semlyen, A. (1999). Rational approximation of frequency domain responses by vector fitting. *IEEE Transactions on power delivery*, 14(3):1052–1061.

Haneda, Y., Makino, S., and Kaneda, Y. (1994). Common acoustical pole and zero modeling of room transfer functions. *IEEE Transactions on Speech and Audio Processing*, 2(2):320–328.

Haneda, Y., Makino, S., Kaneda, Y., and Kitawaki, N. (1999). Common-acoustical-pole and zero modeling of head-related transfer functions. *IEEE Transactions on Speech and Audio Processing*, 7(2):188–196.

Hartmann, W. M. (1983). Localization of sound in rooms. *The Journal of the Acoustical Society of America*, 74(5):1380–1391.

Hill, P. A., Nelson, P. A., Kirkeby, O., and Hamada, H. (2000). Resolution of front–back confusion in virtual acoustic imaging systems. *The Journal of the Acoustical Society of America*, 108(6):2901–2910.

Honda, A., Shibata, H., Gyoba, J., Saitou, K., Iwaya, Y., and Suzuki, Y. (2007). Transfer effects on sound localization performances from playing a virtual three-dimensional auditory game. *Applied Acoustics*, 68(8):885–896.

Hudde, H. and Schmidt, S. (2009). Accuracy of acoustic ear canal impedances: Finite element simulation of measurement methods using a coupling tube. *The Journal of the Acoustical Society of America*, 125(6):3819–3827.

Hugeng, W. W. and Gunawan, D. (2010). Improved Method for Individualization of Head-Related Transfer Functions on Horizontal Plane Using Reduced Number of Anthropometric Measurements. *arXiv preprint arXiv:1005.5137*.

Hwang, S. and Park, Y. (2008). Interpretations on principal components analysis of head-related impulse responses in the median plane. *The Journal of the Acoustical Society of America*, 123(4):EL65–EL71.

Hwang, S., Park, Y., and Park, Y. (2008). Modeling and customization of head-related impulse responses based on general basis functions in time domain. *Acta Acustica united with Acustica*, 94(6):965–980.

Iida, K., Ishii, Y., and Nishioka, S. (2014). Personalization of head-related transfer functions in the median plane based on the anthropometry of the listener's pinnae. *The Journal of the Acoustical Society of America*, 136(1):317–333.

Iida, K., Itoh, M., Itagaki, A., and Morimoto, M. (2007). Median plane localization using a parametric model of the head-related transfer function based on spectral cues. *Applied Acoustics*, 68(8):835–850.

Inoue, N., Kimura, T., Nishino, T., Itou, K., and Takeda, K. (2005). Evaluation of HRTFs estimated using physical features. *Acoustical science and technology*, 26(5):453–455.

International Organization for Standardization (2009). Acoustics - Measurement of Room Acoustic Parameters - Part 1: Performance Spaces. EN ISO 3382-1, Brussels, Belgium.

Iwaya, Y. (2006). Individualization of head-related transfer functions with tournament-style listening test: Listening with other's ears. *Acoustical science and technology*, 27(6):340–343.

Jin, C. T., Guillon, P., Epain, N., Zolfaghari, R., van Schaik, A., Tew, A. I., Hetherington, C., and Thorpe, J. (2014). Creating the Sydney York morphological and acoustic recordings of ears database. *IEEE Transactions on Multimedia*, 16(1):37–46.

Jin, C. T., Leong, P., Leung, J., Corderoy, A., and Carlile, S. (2000). Enabling individualized virtual auditory space using morphological measurements. In IEEE, editor, *Proceedings of the First IEEE Pacific-Rim Conference on Multimedia*.

Jolliffe, I. (2002). *Principal component analysis*. Springer, New York and Berlin and Heidelberg, 2nd edition.

Kahana, Y. and Nelson, P. A. (2007). Boundary element simulations of the transfer function of human heads and baffled pinnae using accurate geometric models. *Journal of sound and vibration*, 300(3):552–579.

Kaneko, S., Suenaga, T., and Fujiwara, M., Kumehara, K., Shirakihara, F., Sekine, S. (2016). Ear Shape Modeling for 3D Audio and Acoustic Virtual Reality: The Shape-Based Average HRTF. In Audio Engineering Society, editor, *61st Conference on Audio for Games*.

Katz, B. F. G. (2001). Boundary element method calculation of individual head-related transfer function. I. Rigid model calculation. *The Journal of the Acoustical Society of America*, 110(5):2440–2448.

Katz, B. F. G. and Noisternig, M. (2014). A comparative study of interaural time delay estimation methods. *The Journal of the Acoustical Society of America*, 135(6):3530–3540.

Katz, B. F. G. and Parseihian, G. (2012). Perceptually based head-related transfer function database optimization. *The Journal of the Acoustical Society of America*, 131(2):EL99–EL105.

Kearney, G. (2015). Spatial Audio for Domestic Interactive Entertainment. http://www.york.ac.uk/sadie-project/binaural.html.

Kirkeby, O., Nelson, P. A., Hamada, H., and Orduna-Bustamante, F. (1998). Fast deconvolution of multichannel systems using regularization. *IEEE Transactions on Speech and Audio Processing*, 6(2):189–194.

Kistler, D. J. and Wightman, F. L. (1992). A model of head–related transfer functions based on principal components analysis and minimum–phase reconstruction. *The Journal of the Acoustical Society of America*, 91:1637.

Klumpp, R. G. and Eady, H. R. (1956). Some measurements of interaural time difference thresholds. *The Journal of the Acoustical Society of America*, 28(5):859–860.

Kuhn, G. F. (1977). Model for the interaural time differences in the azimuthal plane. *The Journal of the Acoustical Society of America*, 62:157.

Kulkarni, A. and Colburn, H. S. (2004). Infinite-impulse-response models of the head-related transfer function. *The Journal of the Acoustical Society of America*, 115(4):1714–1728.

Kulkarni, A., Isabelle, S. K., and Colburn, H. S. (1999). Sensitivity of human subjects to head-related transfer-function phase spectra. *The Journal of the Acoustical Society of America*, 105(5):2821–2840.

Kuttruff, H. (2000). *Room acoustics*. Spon Press, London, 4th edition.

Laakso, T. I., Valimaki, V., Karjalainen, M., and Laine, U. K. (1996). Splitting the unit delay. *Signal Processing Magazine, IEEE*, 13(1):30–60.

Larcher, V. and Jot, J.-M. (1997). Techniques d'interpolation de filtres audio-numériques. In Société Française d'Acoustique, editor, *Proceedings du Congres Français d'Acoustique*.

Leishman, T. W., Rollins, S., and Smith, H. M. (2006). An experimental evaluation of regular polyhedron loudspeakers as omnidirectional sources of sound. *The Journal of the Acoustical Society of America*, 120(3):1411–1422.

Lindau, A., Estrella, J., and Weinzierl, S. (2010). Individualization of dynamic binaural synthesis by real time manipulation of ITD. In Audio Engineering Society, editor, *128th Audio Engineering Society Convention*.

Lins, M., Bomhardt, R., and Fels, J. (2016). Individualisierung der HRTF: Ein Ellipsoidmodell zur Anpassung von interauralen Pegeldifferenzen. In Deutsche Gesellschaft für Akustik, editor, *42. Jahrestagung für Akustik*.

Lopez-Poveda, E. A. and Meddis, R. (1996). A physical model of sound diffraction and reflections in the human concha. *The Journal of the Acoustical Society of America*, 100:3248.

Maazaoui, M. and Warusfel, O. (2016). Estimation of Individualized HRTF in Unsupervised Conditions. In Audio Engineering Society, editor, *140th International AES Convention*.

Macpherson, E. A. and Middlebrooks, J. C. (2002). Listener weighting of cues for lateral angle: the duplex theory of sound localization revisited. *The Journal of the Acoustical Society of America*, 111(5):2219–2236.

Majdak, P., Balazs, P., and Laback, B. (2007). Multiple exponential sweep method for fast measurement of head-related transfer functions. *Journal of the Audio Engineering Society*, 55(7/8):623–637.

Majdak, P., Masiero, B., and Fels, J. (2013). Sound localization in individualized and non-individualized crosstalk cancellation systems. *The Journal of the Acoustical Society of America*, 133(4):2055–2068.

Makous, J. C. and Middlebrooks, J. C. (1990). Two–dimensional sound localization by human listeners. *The Journal of the Acoustical Society of America*, 87(5):2188–2200.

Masiero, B. (2012). *Individualized binaural technology: measurement, equalization and perceptual evaluation*. Phd thesis, RWTH Aachen University, Aachen.

Masiero, B. and Fels, J. (2011). Perceptually robust headphone equalization for binaural reproduction. In Audio Engineering Society, editor, *130th Audio Engineering Society Convention*.

Masiero, B., Pollow, M., and Fels, J. (2011). Design of a fast broadband individual head-related transfer function measurement system. In European Acoustics Association, editor, *6th Forum Acusticum* .

Mason, R., Ford, N., Rumsey, F., and de Bruyn, B. (2001). Verbal and nonverbal elicitation techniques in the subjective assessment of spatial sound reproduction. *Journal of the Audio Engineering Society*, 49(5):366–384.

McFadden, D. (1998). Sex differences in the auditory system. *Developmental Neuropsychology*, 14(2-3):261–298.

McFadden, D. and Pasanen, E. G. (1976). Lateralization at high frequencies based on interaural time differences. *The Journal of the Acoustical Society of America*, 59(3):634–639.

McMullen, K., Roginska, A., and Wakefield, G. (2012). Subjective selection of head-related transfer functions (HRTFs) based on spectral coloration and interaural time differences (ITD) cues. In Audio Engineering Society, editor, *133rd Audio Engineering Society Convention*.

Mechel, F. P. (2008). *Formulas of acoustics*. Springer, Berlin and London, 2nd edition.

Mehrgardt, S. and Mellert, V. (1977). Transformation characteristics of the external human ear. *The Journal of the Acoustical Society of America*, 61(6):1567–1576.

Middlebrooks, J. C. (1999a). Individual differences in external-ear transfer functions reduced by scaling in frequency. *The Journal of the Acoustical Society of America*, 106:1480.

Middlebrooks, J. C. (1999b). Virtual localization improved by scaling nonindividualized external-ear transfer functions in frequency. *The Journal of the Acoustical Society of America*, 106:1493.

Middlebrooks, J. C. and Green, D. M. (1990). Directional dependence of interaural envelope delays. *The Journal of the Acoustical Society of America*, 87:2149.

Middlebrooks, J. C. and Green, D. M. (1991). Sound localization by human listeners. *Annual review of psychology*, 42(1):135–159.

Middlebrooks, J. C. and Green, D. M. (1992). Observations on a principal components analysis of head–related transfer functions. *The Journal of the Acoustical Society of America*, 92(1):597–599.

Mills, A. W. (1958). On the minimum audible angle. *The Journal of the Acoustical Society of America*, 30:237.

Mills, A. W. (1960). Lateralization of High–Frequency Tones. *The Journal of the Acoustical Society of America*, 32(1):132–134.

Mokhtari, P., Takemoto, H., Nishimura, R., and Kato, H. (2015). Frequency and amplitude estimation of the first peak of head-related transfer functions from individual pinna anthropometry. *The Journal of the Acoustical Society of America*, 137(2):690–701.

Møller, H., Hammershøi, D., Jensen, C. B., and Sørensen, M. F. (1995a). Transfer characteristics of headphones measured on human ears. *Journal of the Audio Engineering Society*, 43(4):203–217.

Møller, H., Sørensen, M. F., Hammershøi, D., and Jensen, C. B. (1995b). Head-related transfer functions of human subjects. *Journal of the Audio Engineering Society*, 43(5):300–321.

Morimoto, M. and Aokata, H. (1984). Localization cues of sound sources in the upper hemisphere. *Journal of the Acoustical Society of Japan (E)*, 5(3):165–173.

Nam, J., Kolarm, M. A., and Abel, J. S. (2008). On the minimum-phase nature of head-related transfer functions. In Audio Engineering Society, editor, *125th AES Convention*.

Nishino, T., Inoue, N., Takeda, K., and Itakura, F. (2007). Estimation of HRTFs on the horizontal plane using physical features. *Applied Acoustics*, 68(8):897–908.

Ohm, J.-R. and Lüke, H. D. (2010). *Signalübertragung: Grundlagen der digitalen und analogen Nachrichtenübertragungssysteme.* Springer, Berlin and Heidelberg, 11th edition.

Oppenheim, A. V., Schafer, R. W., and Buck, J. R. (1999). *Discrete-time signal processing.* Prentice Hall signal processing series. Prentice Hall, Upper Saddle River, N.J, 2nd edition.

Parseihian, G. and Katz, B. F. G. (2012). Rapid head-related transfer function adaptation using a virtual auditory environment. *The Journal of the Acoustical Society of America,* 131(4):2948–2957.

Perrett, S. and Noble, W. (1997). The effect of head rotations on vertical plane sound localization. *The Journal of the Acoustical Society of America,* 102(4):2325–2332.

Perrott, D. R. (1969). Role of signal onset in sound localization. *The Journal of the Acoustical Society of America,* 45(2):436–445.

Phillips, D. P. and Hall, S. E. (2005). Psychophysical evidence for adaptation of central auditory processors for interaural differences in time and level. *Hearing research,* 202(1):188–199.

Pikler, A. G. (1966). Logarithmic frequency systems. *The Journal of the Acoustical Society of America,* 39(6):1102–1110.

Pollack, I. and Rose, M. (1967). Effect of head movement on the localization of sounds in the equatorial plane. *Attention, Perception, & Psychophysics,* 2(12):591–596.

Pollow, M., Nguyen, K.-V., Warusfel, O., Carpentier, T., Müller-Trapet, M., Vorländer, M., and Noisternig, M. (2012). Calculation of head-related transfer functions for arbitrary field points using spherical harmonics decomposition. *Acta Acustica united with Acustica,* 98(1):72–82.

Rafaely, B. (2005). Analysis and design of spherical microphone arrays. *IEEE Transactions on Speech and Audio Processing,* 13(1):135–143.

Ramos, O. and Tommansini, F. (2014). Magnitude Modelling of HRTF Using Principal Component Analysis Applied to Complex Values. *Archives of Acoustics,* 39(4):477–482.

Rasumow, E., Blau, M., Hansen, M., van de Par, S., Doclo, S., Mellert, V., and Püschel, D. (2014). Smoothing individual head-related transfer functions in the frequency and spatial domains. *The Journal of the Acoustical Society of America,* 135(4):2012–2025.

Raykar, V. C., Duraiswami, R., and Yegnanarayana, B. (2005). Extracting the frequencies of the pinna spectral notches in measured head related impulse responses. *The Journal of the Acoustical Society of America,* 118:364–374.

Rayleigh, L. (1907). XII. On our perception of sound direction. *The London, Edinburgh, and Dublin Philosophical Magazine and Journal of Science,* 13(74):214–232.

Richter, J.-G., Behler, G., and Fels, J. (2016). Evaluation of a Fast HRTF Measurement System. In Audio Engineering Society, editor, *140th International AES Convention*.

Romigh, G. D., Brungart, D. S., Stern, R. M., and Simpson, B. D. (2015). Efficient real spherical harmonic representation of head-related transfer functions. *IEEE Journal of Selected Topics in Signal Processing*, 9(5):921–930.

Runkle, P., Yendiki, A., and Wakefield, G. (2000). Active sensory tuning for immersive spatialized audio. In International Community for Auditory Display, editor, *International Conference on Auditory Display*.

Satarzadeh, P., Algazi, V. R., and Duda, R. O. (2007). Physical and filter pinna models based on anthropometry. In Audio Engineering Society, editor, *122nd Audio Engineering Society Convention*.

Savioja, L., Huopaniemi, J., Lokki, T., and Väänänen, R. (1999). Creating interactive virtual acoustic environments. *Journal of the Audio Engineering Society*, 47(9):675–705.

Schmitz, A. and Bietz, H. (1998). Free-Field Diffuse-Field Transformation of Artificial Heads. In Audio Engineering Society, editor, *105th AES Convention*, volume 105.

Schönstein, D. and Katz, B. F. G. (2010). HRTF selection for binaural synthesis from a database using morphological parameters. In Australian Acoustical Society, editor, *20th International Congress on Acoustics*.

Seber, G. A. F and Lee, A. J. (2003). *Linear regression analysis*. Wiley series in probability and statistics. Wiley-Interscience, Hoboken, N.J., 2nd edition.

Seeber, B. and Fastl, H. (2003). Subjective selection of non-individual head-related transfer functions. In International Community for Auditory Display, editor, *International Conference on Auditory Display*.

Senova, M. A., McAnally, K. I., and Martin, R. L. (2002). Localization of virtual sound as a function of head-related impulse response duration. *Journal of the Audio Engineering Society*, 50(1/2):57–66.

Shaw, E. A. G. and Teranishi, R. (1968). Sound Pressure Generated in an External–Ear Replica and Real Human Ears by a Nearby Point Source. *The Journal of the Acoustical Society of America*, 44(1):240–249.

Shin, K. and Park, Y. (2008). Enhanced vertical perception through head-related impulse response customization based on pinna response tuning in the median plane. *IEICE Transactions on Fundamentals of Electronics, Communications and Computer Sciences*, 91(1):345–356.

Shinn-Cunningham, B. G. (2001). Localizing sound in rooms. *ACM/SIGGRAPH and Eurographics Campfire: Acoustic Rendering for Virtual Environments*, pages 1–6.

Shinn-Cunningham, B. G., Durlach, N. I., and Held, R. M. (1998). Adapting to supernormal auditory localization cues. I. Bias and resolution. *The Journal of the Acoustical Society of America*, 103(6):3656–3666.

Shinn-Cunningham, B. G., Kopco, N., and Martin, T. J. (2005). Localizing nearby sound sources in a classroom: Binaural room impulse responses. *The Journal of the Acoustical Society of America*, 117(5):3100–3115.

Silzle, A. (2002). Selection and tuning of HRTFs. In Audio Engineering Society, editor, *112nd Audio Engineering Society Convention*.

Simon, L. S. R., Andreopoulou, A., and Katz, B. F. G. (2016). Investigation of Perceptual Interaural Time Difference Evaluation Protocols in a Binaural Context. *Acta Acustica united with Acustica*, 102(1):129–140.

Sottek, R. and Genuit, K. (1999). Physical modeling of individual head-related transfer functions. In Deutsche Gesellschaft für Akustik, editor, *Tagung der Deutschen Arbeitsgemeinschaft für Akustik*.

Spagnol, S. and Geromazzo, M. (2010). Estimation and modeling of pinna-related transfer functions. In DAFx, editor, *13th International Conference on Digital Audio Effects*.

Spagnol, S., Geronazzo, M., and Avanzini, F. (2013). On the relation between pinna reflection patterns and head-related transfer function features. *Audio, Speech, and Language Processing, IEEE Transactions on*, 21(3):508–519.

Spagnol, S. and Hiipakka, M. (2011). A single-azimuth pinna-related transfer function database. In DAFx, editor, *14th International Conference on Digital Audio Effects*.

Takemoto, H., Mokhtari, P., Kato, H., Nishimura, R., and Iida, K. (2012). Mechanism for generating peaks and notches of head-related transfer functions in the median plane. *The Journal of the Acoustical Society of America*, 132(6):3832–3841.

Thompson, L. L. and Pinsky, P. M. (1994). Complex wavenumber Fourier analysis of the p-version finite element method. *Computational Mechanics*, 13(4):255–275.

Tobias, J. V. and Schubert, E. D. (1959). Effective onset duration of auditory stimuli. *The Journal of the Acoustical Society of America*, 31(12):1595–1605.

Tohyama, M., Lyon, R. H., and Koike, T. (1994). Phase variabilities and zeros in a reverberant transfer function. *The Journal of the Acoustical Society of America*, 95(1):286–296.

Torres-Gallegos, E. A., Orduna-Bustamante, F., and Arámbula-Cosío, F. (2015). Personalization of head-related transfer functions (HRTF) based on automatic photo-anthropometry and inference from a database. *Applied Acoustics*, 97:84–95.

Tribolet, J. (1977). A new phase unwrapping algorithm. *IEEE Transactions on Acoustics, Speech, and Signal Processing*, 25(2):170–177.

Wallach, H. (1940). The role of head movements and vestibular and visual cues in sound localization. *Journal of Experimental Psychology*, 27(4):339.

Warusfel, O. (2002). Listen HRTF database. http://recherche.ircam.fr/equipes/salles/listen/.

Watanabe, K., Iwaya, Y., Suzuki, Y., Takane, S., and Sato, S. (2014). Dataset of head-related transfer functions measured with a circular loudspeaker array. *Acoustical science and technology*, 35(3):159–165.

Watanabe, K., Ozawa, K., Iwaya, Y., Suzuki, Y., and Aso, K. (2007). Estimation of interaural level difference based on anthropometry and its effect on sound localization. *The Journal of the Acoustical Society of America*, 122:2832.

Watson, A. B. and Pelli, D. G. (1983). QUEST: A Bayesian adaptive psychometric method. *Perception & psychophysics*, 33(2):113–120.

Wenzel, E. M., Arruda, M., Kistler, D. J., and Wightman, F. L. (1993). Localization using nonindividualized head–related transfer functions. *The Journal of the Acoustical Society of America*, 94:111.

Wightman, F. L. and Kistler, D. J. (1989). Headphone simulation of free–field listening. I: Stimulus synthesis. *The Journal of the Acoustical Society of America*, 85(2):858–867.

Wightman, F. L. and Kistler, D. J. (1992). The dominant role of low–frequency interaural time differences in sound localization. *The Journal of the Acoustical Society of America*, 91:1648–1661.

Wightman, F. L. and Kistler, D. J. (1993). Multidimensional scaling analysis of head-related transfer functions. In IEEE, editor, *1993 IEEE Workshop on Applications of Signal Processing to Audio and Acoustics*.

Wightman, F. L. and Kistler, D. J. (1999). Resolution of front–back ambiguity in spatial hearing by listener and source movement. *The Journal of the Acoustical Society of America*, 105(5):2841–2853.

Wightman, F. L. and Kistler, D. J. (2005). Measurement and validation of human HRTFs for use in hearing research. *Acta Acustica united with Acustica*, 91(3):429–439.

Woodworth, R. S. (1940). Experimental psychology. *The Journal of Nervous and Mental Disease*, 91(6):811.

Xie, B.-S. (2012). Recovery of individual head-related transfer functions from a small set of measurements a. *The Journal of the Acoustical Society of America*, 132(1):282–294.

Xie, B.-S. and Zhang, T. (2010). The audibility of spectral detail of head-related transfer functions at high frequency. *Acta Acustica united with Acustica*, 96(2):328–339.

Xu, S., Li, Z., and Salvendy, G. (2008). Individualized head-related transfer functions based on population grouping. *The Journal of the Acoustical Society of America*, 124(5):2708–2710.

Zhong, X.-L. and Xie, B.-S. (2007). A novel model of interaural time difference based on spatial Fourier analysis. *Chinese Physics Letters*, 24(5):1313.

Zhong, X.-L. and Xie, B.-S. (2013a). An individualized interaural time difference model based on spherical harmonic function expansion. *Chinese Journal of Acoustics*, 3:10.

Zhong, X.-L. and Xie, B.-S. (2013b). Consistency among the head-related transfer functions from different measurements. In Acoustical Society of America, editor, *21th International Congress on Acoustics*, volume 19.

Zhong, X.-L., Zhang, F.-C., and Xie, B.-S. (2013). On the spatial symmetry of head-related transfer functions. *Applied Acoustics*, 74(6):856–864.

Ziegelwanger, H. and Majdak, P. (2014). Modeling the direction-continuous time-of-arrival in head-related transfer functions. *The Journal of the Acoustical Society of America*, 135(3):1278–1293.

Ziegelwanger, H., Majdak, P., and Kreuzer, W. (2015). Numerical calculation of listener-specific head-related transfer functions and sound localization: Microphone model and mesh discretization. *The Journal of the Acoustical Society of America*, 138(1):208–222.

Zotkin, D. N., Duraiswami, R., Grassi, E., and Gumerov, N. A. (2006). Fast head-related transfer function measurement via reciprocity. *The Journal of the Acoustical Society of America*, 120(4):2202–2215.

Zotkin, D. N., Hwaiig, J., Duraiswami, R., and Davis, L. S. (2003). HRTF personalization using anthropometric measurements. In IEEE, editor, *2003 IEEE Workshop on the Applications of Signal Processing to Audio and Acoustics*.

Zündorf, I. C., Karnath, H.-O., and Lewald, J. (2011). Male advantage in sound localization at cocktail parties. *Cortex*, 47(6):741–749.

Zwislocki, J. and Feldman, R. S. (1956). Just noticeable differences in dichotic phase. *The Journal of the Acoustical Society of America*, 28(5):860–864.

Curriculum Vitae

Personal Data

Name	Ramona Bomhardt
17.09.1984	born in Hess. Lichtenau, Germany

Educational Background

1991–1997	Karlheinz-Böhm-Schule Waldkappel, Germany
1997–2004	Freiherr-vom Stein-Schule Hess. Lichtenau, Germany

Higher Education

2004–2010	Diploma Studies in Electrical Engineering and Information Technology RWTH Aachen University, Germany
	Internship at Antennentechnik Bad Blankenburg Bad Blankenburg, Germany

Professional Experience

2010–2011	Vehicle NVH Project Engineer Bertrandt, Cologne, Germany
2011–2017	Research Assistant, Institute of Technical Acoustics RWTH Aachen University, Germany

Aachen, Germany, August 14, 2017

Danksagung

Meine Promotion war nicht nur durch meine eigene Forschungsarbeit möglich, sondern zu mindestens gleichen Teilen durch die Unterstützung und Motivation vieler netter Menschen in meinem Umfeld. Aus diesem Grund möchte ich diesen Personen explizit meinen Dank aussprechen.

Zuerst möchte ich Prof. Janina Fels danken, dass sie mir die Promotion ermöglicht und mich über die Jahre am Institut betreut hat. Prof. Peter Jax, der sich als Zweitgutachter bereiterklärt hat und an entsprechenden Stellen der Arbeit nachzuhaken, gilt ebenfalls mein Dank.

Die gute Atmosphäre im Institut trug stets zur Motivation bei der Arbeit bei und lieferte entscheidende Impulse während der Arbeit. Aus diesem Grund möchte ich mich bei allen Mitarbeitern des Instituts für Technische Akustik bedanken. Hier haben mir insbesondere Diskussionen mit Dr. Pascal Dietrich, Dr. Martin Guski, Marcia Lins und Dr. Frank Wefers neue mathematische und signaltheoretische Dimensionen aufgespannt. Des Weiteren bin ich für das stets offene Ohr und die hilfreichen Hinweise von Prof. Michael Vorländer und Dr. Gottfried Behler sehr dankbar. An dieser Stelle sind außerdem die Werkstätten von Uwe Schlömer und Rolf Kaldenbach zu nennen, die mir mit ihrer Kompetenz zur Seite standen. Neben den Mitarbeitern haben Philipp Bechtel, Hark Braren und Armin Erraji deutlich zum Vorankommen meiner Arbeit beigetragen. Die Bürokratie konnte ich während meiner Zeit am Institut durch die helfende Hand von Karin Charlier effizient bewältigen, sodass ich mich stärker auf die wissenschaftliche Arbeit konzentrieren konnte. Den Ausgleich zum Institutsalltag – bei dem es trotzdem gelegentlich zu fachlichen Diskussionen kam – lieferten Marco Berzborn, Johannes Klein, Michael Kohnen und Rob Opdam beim Klettern, Laufen oder Ski fahren.

Nicht nur das Institut, sondern auch meine Freunde haben mich während der letzten Jahre begleitet. Bei Ihnen möchte ich mich dafür bedanken, dass ich immer auf sie zählen kann. Ich bin sehr stolz darauf, dass meine Eltern immer zu mir stehen und ohne die der Weg bis zur Promotion nicht möglich gewesen wäre. Unglaublich dankbar bin ich Christian Glatzer, der mir Halt und seine Liebe schenkt.

Bisher erschienene Bände der Reihe
Aachener Beiträge zur Akustik

ISSN 1866-3052
ISSN 2512-6008 (seit Band 28)

Alle erschienenen Bücher können unter der angegebenen ISBN-Nummer direkt online (http://www.logos-verlag.de) oder per Fax (030 - 42 85 10 92) beim Logos Verlag Berlin bestellt werden.